TOWARD A NEW DIMENSION

TOWARD A NEW DIMENSION

Exploring the Nanoscale

ANNE MARCOVICH
and
TERRY SHINN

UNIVERSITY PRESS

Great Clarendon Street, Oxford, OX2 6DP,
United Kingdom

Oxford University Press is a department of the University of Oxford.
It furthers the University's objective of excellence in research, scholarship,
and education by publishing worldwide. Oxford is a registered trade mark of
Oxford University Press in the UK and in certain other countries

First Edition published in 2014

Reprinted 2016

Impression: 2

Published in the United States of America by Oxford University Press
198 Madison Avenue, New York, NY 10016, United States of America

British Library Cataloguing in Publication Data
Data available

Library of Congress Control Number: 2014931557

ISBN 978–0–19–871461–3

Printed in Great Britain by
CPI Group (UK) Ltd, Croydon, CR0 4YY

For Paul Forman—a loyal and wise friend and exceptional scholar and colleague.
– Terry Shinn

For Jonathan, Esther, and Judith Elbaz—for their solidarity throughout this long project and for their love.
– Anne Marcovich

PREFACE

The research and writing of this book constitute one piece of our long-standing reflection about the theme of transformation in contemporary scientific research. It is commonly acknowledged that science has changed considerably over the course of the last century. In turning our gaze to research on the nanoscale, we asked what does the birth and growth of nanoscience tell us about the evolution of research? At the opening of this study, we were confronted by two contradictory answers to our questions about the transformative character of nanoresearch: "no, with nano there is nothing new under the Sun," and "yes, in nano, scientific investigation is being modified in many important ways." In view of these responses, nanoscale research appeared to be a privileged window to explore the confluence of the old and the new. The scientists' answers referred to the fact that science's theoretical foundations had not been seriously recast during the last decades, but that research practice and epistemology have become in important respects significantly different. Nanoscale research is a child of material, technological, and epistemological novelties. But at the same time, aspects of reasoning dominant during the nineteenth century, that had declined across much of the twentieth century, have suddenly and surprisingly been rekindled in combination with emerging technologies and accompanying experimental practices.

The epistemological arrow that historically slaved research questions to the materials available in nature has been reversed in nanoscale scientific investigation. This has opened the path for new ways of thinking and doing. Indeed since the 1980s, nanoscale research has been capable of synthesizing pre-designed materials, atom by atom and molecule by molecule, making it possible for experimenters to exactingly tailor the material objects to which they direct their questions. This constitutes a decisive historical turning point. Nanoscale research is the landscape of material control. Inside nano, an entirely new specialty has emerged that is devoted to the design, synthesis, and fabrication of preconceived objects.

Nanoscale research is inextricably linked to the genesis of a new species of metrological instrumentation. A category of instruments termed scanning probe microscopy lies at the heart of this material revolution. Such instrumentation can study (see) single molecules and atoms and can also shift them about and even attach them one to another, at will. A second recently introduced category of devices, "computational instrumentation," often known as "numerical simulation," has pervaded the research universe since the 1990s; but nowhere more so than in the area of nanoscale studies. Simulation based investigation has in several cases designed and explored nanostructured materials even prior to their concrete materialization. The relationship between metrological and computational experimentation in the nano field constitutes a deeply ingrained combinatorial that is central to nano culture. Scanning probe microscopy and numerical simulation both express information about the structure of nanoscale objects in terms of visual images. They simultaneously provide a synthetic and local picture of phenomena. For example, by sharpening pictorial contrasts of structure and by introducing differentiating color schemes into research images, practitioners painstakingly generate a cartography and corresponding comprehension of often nuanced nanoscale properties and processes. Of utmost importance, images operate as a common language between metrologists and simulators, thus promoting complementarity. They are indeed an epistemological cornerstone of nanoscience.

The function of images is tightly connected with questions of form. Images as drawings and also photographs were for a long time—throughout the nineteenth century and even before—important in science prior to nano. This was certainly the case with reference to things geographic, geological, botanic, zoological, anatomical, crystallographic, etc. It is essential to grasp that the pivotal position of form in nano explicitly constitutes a kind of renaissance of an earlier historical era. The morphology of objects and sometimes of their behavior is a key question in nanoscale research, thus giving a renewed importance to the traditional epistemological parameter of form. Through the new instrumentation of imaging, form has recovered its earlier importance that during much of the twentieth century had subsequently often been supplanted by mathematical formalism or statistical representations.

Taken together, images and form reinject an epistemology of descriptivism which was central to much pre-twentieth-century science. Descriptivism is a way of seeing the world, reflecting about it, ascribing relationships and explanation, and finally communicating about the world. In nano, description focuses on the highly local and particular. Nanometric landscapes are worthy of description in and of themselves, and not least of all for their local features. The combinatorial of description and local privileges the unique as opposed to the highly general. It is partly for this reason that theory and models are not the main expectation of nanoscientists.

Since the end of the nineteenth century, the microscopic universe, the worlds of molecules and atoms, was principally long described and explained in terms of statistical probability—the indeterminacy approach. This is certainly the case with the Schrödinger equation and the like. Scientists were restricted to determining the probability of an event. However, with the advent of nanoscale scientific research, with its battery of novel instrumentation and materials, under specified conditions it is now common practice to identify the position of a single molecule and even to study the morphology of its surface, as Galileo studied the landscape of the moon gazing through his telescope some four hundred years ago. Molecules and atoms are now understood in terms of a determinist epistemology, as opposed to a stochastic, probabilistic epistemology. The very action of molecular and atomic control constitutes the materialization of determinism.

The reader can see, therefore, that the nanoscale scientific enterprise rejuvenates many older intellectual traditions as they are re-introduced through new instrumentation and material control. We thus no longer frame the issue of transformations of contemporary scientific research in terms of *"no, there is nothing new under the Sun,"* versus *"yes, scientific investigation is being renewed in many important ways,"* but now we reason in terms of subtle combinatorials between the perpetuation of nineteenth-century epistemology and late twentieth-century materials and instrument evolution. It is this peculiar balance that allows us to effectively address the question of complex change in contemporary scientific research.

The issue of the significance of what has emerged in the course of recent scientific investigation, refers not only to matters of instrumentation, research-objects, and epistemology, but also extends to the question of the organization and institutions of research. The debate between the discipline based structuring of science versus interdisciplinarity continues to rage. Proponents of the disciplinary organization of work insist that original investigation must be tightly defined and that it occurs best in a disciplinary matrix. By contrast, pro-interdisciplinary advocates insist on cognitive complexity, heterogeneity, and fluidity, and they claim that innovative progress is not consistent with a disciplinary frame. Our investigation of nanoscience points in the direction of disciplinary structures, but that are reshaped in the context of contemporary cognitive practices which often entail well-defined, temporary collaborations of practitioners from different stable, established disciplinary domains. At the birth of nanoscale research, some practitioners predicted that the field would ultimately emerge as a new, autonomous scientific discipline. In contrast and more recently, many observers have argued that nano is totally fragmented, where each specialty possesses its specific nano routines.

Our study of nanoscale scientific research instead suggests that a "transversalist" perspective of nanoscience, and perhaps beyond nano (for many other areas of science), may offer a more perspicacious and balanced vision. As seen from our proposed transversalist perspective, nanoscale practitioners, whatever their home field, can work inside the framework of their homeland discipline while simultaneously speaking across their respective borderland to colleagues working in other fields. Boundaries and circumscribed circulation are here compatible. We refer to this species of organizational, institutional, cognitive configuration as the "new disciplinarity." Here, the maintenance of strong disciplinary boundaries does not preclude a practitioner standing on their disciplinary borderland and shouting across the boundary wall to colleagues who too work in their home disciplines and who also shout over the separating wall! If our assessment is correct, then the concept and activities of transversality, as vehicled in the new disciplinarity, are slated to become a component in the transformational processes of contemporary science—still another possible message gleaned from exploring the nanoscale.

ACKNOWLEDGEMENTS

We wish to thank the scholars who generously read our manuscript and provided commentary in the capacity of colleagues or reviewers: Bernadette Bensaud-Vincent (University of Paris I Sorbonne, Maria Caramez-Carlotto (University of Sao Paulo), Johannes Lenhard (Bielefeld University), Cyrus Moody (Rice University), Alfred Nordmann (Darmstadt University), Arnaud de Saint-Martin (Centre National de la Recherche Scientifique, Paris). Among the many scientists who we interviewed during our research, we are particularly indebted to: Tristan Cren, Roger Grousson, Bernard Jusserand, Claudine Noguera, Bernard Perrin, Valia Voliotis (Institut des Nanosciences, Paris), Gerald Dujardin, Philippe Minard, (University of Orsay, France), Uzi Landman (Georgia Institute of Technology), Paul Rothemund (Caltech University), Ned Seeman (New York University), Shimon Weiss (University of UCLA). We also thank Alexandra Frenod (Groupe d'Etude des Methodes de l'Analyse Sociologique de la Sorbonne, Centre National de la Recherche Scientifique) and Nora Scott for their alert editing of our manuscript. We are also indebted to Matthieu Renard, the art graphist who designed the book cover. Finally, it has been a pleasure to deal with Sonke Adlung and Jessica White at Oxford University Press who have constantly proven efficient and kind in the birth of the present volume.

CONTENTS

LIST OF ABBREVIATIONS

AFM	atomic force microscope
CLSM	confocal laser scanning microscope
CVD	chemical vapor deposition
DFT	density functional theory
ESQC	elastic scattering quantum chemistry
FESEM	field-emission scanning electron microscopy
FRET	Förster resonance electron transfer or fluorescent resonance electron transfer
FTIR	Fourier transform infra-red spectroscopy
LEED	Low energy electron diffraction
MBE	molecular beam epitaxy
NEMS	nanoelectromechanical systems
NFOM	near field optical microscopy
NNI	nanotechnology initiative
NMR	nuclear magnetic resonance
NSR	nanoscale research
NST	nanoscience and technology
OES	optical emission spectroscopy
PCR	polymerase chain reaction
PES	photo-electron spectroscopy
QD	quantum dots
SAMs	self-assembling monolayers
SEM	scanning electron microscope
SPM	scanning probe microscopy
STM	scanning tunneling microscope
STS	scanning tunneling spectroscopy
TEM	transmission electron microscope
VR	virtual reality
XPS	X-ray photo-electron microscopy

Introduction

Is the world of nano not in some ways akin to Galileo turning his telescope inward toward the nano universe? Galileo observed the landscape of the Moon and the satellites of Jupiter. Has nano not also "looked with the eye" onto the topography of single molecules and the geography of atomic constellation? On the other hand, Galileo, Copernicus, and Kepler could not conceive of rearranging the configuration and material of the Sun's planetary system in order to test and thus to better grasp the laws of nature. Yet nano can for its part modify or even create materials in such a way that scientists introduce artificial conditions by designing and fabricating molecule-scale universes for dreamed-of experimental study of objects that do not exist in nature. Through design, control, and observation on the scale of single molecules and atoms, nano replaces, or at least complements, the dominant long-standing probabilistic quantum perception of the micro universe with a deterministic epistemology of objects as stable forms; and with it, it correspondingly introduces an alternative neo-descriptivist paradigm.

In the land of nano-investigation, where the dimensions of objects neighbors a billionth of a meter, science has discovered that physical properties are determined strongly by size and form. Here the learning derived from the century-old exploration of "bulk materials," whose dimensions range between a few hundred nanometers and infinity, are no longer relevant. Are the introduction, evolution, and substance of this new way of observing, doing, and thinking not worthy of historical, epistemological, and sociological attention?

The orientation of this book contrasts with the direction of much contemporary scholarship, which examines scientific research in terms of the public understanding of science, government research policy, innovation, communication, and the evolution of the institution of science in post-modern culture. Our study instead focuses on intra-cognitive elements: we investigate the place of instrumentation and materials in the origins and structuring of nanoscale research (or NSR) and explore questions of epistemology with reference to form, image, descriptivism, and determinism. The intra-cognitive side of science is investigated as opposed to an alternative path which examines its extra-cognitive dimensions. In short, this book stands at the crossroads of the history of ideas in contemporary science and its sociology.

The rapid expansion and considerable cognitive achievements of research on nanodimension objects spring from two factors: (1) the ability to observe and analyze the surface features of single molecules and their position; (2) the power to control their spatial relations. The originality of nanoscale research and its distinction from other fields of science revolve around this tandem. The connection between single molecules and control notably endows nanoscale research with a measure of transversality of scientific relevance. Single-molecule observation and atom-by-atom and molecular-based designed, made-to-order materials are today central to the growth of contemporary knowledge in both the physical sciences and the life sciences. This transversality of nanoscale investigation runs counter to the received view that nano constitutes an acutely fragmented field. On a different register, the physical characteristics of matter have traditionally defined and restricted the questions that can be formulated in a research project. In the case of nano-investigations, however, the relationship between material and question is reversed. Why is this? The power to design and build novel materials signifies that questions can now fuel materials rather than being slaves to them. This is a foremost contribution of science nano style.

In nanoscale research, the observational empowerment acquired through the invention of the scanning tunneling microscope, or STM (1981), and of its cousin the atomic force microscope, or AFM (1985), has now made it possible for the first time in the annals of research to explore single molecules, to study an individual molecule's surface features and internal structure. These instruments also enable nanopractitioners to identify the location of an atom, to observe the sometimes complex and surprising geometry of interfacing or locking. In addition to the possibility of dealing with single objects, in nanoscale research, the capacity to design and to synthesize an expanding variety of man-made, artificial materials are also synonymous with nanoscale science. Materials include fullerenes, and in particular carbon nanotubes, and equally decisive, although perhaps less well known, low-dimensional materials such as nanowires, nanowells, and quantum dots. The crucial point here is that these novel materials and instruments now open the way to an unprecedented dialog between the articulation of research questions and what we term "materials by design." Note that objects of nanometric dimensions exhibit properties or behaviors absent or different from bulk materials. Bulk materials are particles whose dimensions exceed one micron, and their surface is small compared with their volume. In nano objects, large surface-to-mass ratios transform physical characteristics.

Recall that a nanometer measures one billionth of a meter; this is equivalent to one ten thousandth of the diameter of a human hair! Precise control at this scale is now routine. Control is situated at the intersection of instrumentation and material. For example, devices like the scanning tunneling microscope enable the manipulation of atoms so precisely that words can be spelled out using clusters of individual atoms to form letters. This was demonstrated in 1989, in a famous article by Donald Eigler in which he wrote the abbreviation IBM with 35 xenon atoms. This feat astounded scientists the world over and alerted them to the latent promise of nanoscale research. Such exact manipulation is now common practice; it is regularly mobilized to create defined molecular and atomic architectures that are the foundation of unprecedented materials, and it underpins experimental research on them. These materials are the bearers of previously unstudied and even unknown physical phenomena.

The scanning tunneling microscope and related instruments are devices that detect the presence of molecular and atomic objects, and identify form, surface features, and inter-molecule interfacing. They are measurement devices—"metrological" instruments. At the time of their inception and initial introduction in research in the 1980s, another species of instrumentation, namely numerical simulation, was increasingly penetrating scientific research. Simulation employs algorithms and models to explore properties of mathematically expressed phenomena and events. In nanoscale research, metrological and numerical instrumentation grew up hand in hand. They are remarkably intertwined. In the following pages, we refer to this interlacing relationship as a "combinatorial." More generally, combinatorial refers to a reciprocal interaction between any two or more entities, be they instrumental, material, phenomenological, or cognitive. Combinatorials abound in nanoscale research to such an extent that they constitute one of the domain signatures. The above-remarked link between instruments and materials synthesis comprises another noteworthy combinatorial, as it contributes to nanoscale research through the multiplication and constant circulation of novel substances and the unending dilation of experimentation horizons.

The birth and development of nanoscale research has also been accompanied by the introduction of several epistemology related paths: form, descriptivism, and determinism. NSR is largely a world where individual objects (molecules and atoms) are studied in terms of their morphology and relative position. For roughly the last two hundred years, scientists have known that matter consists of tiny particles, which could only be studied in terms of their effects. Previously, scientists mainly saw tracks of these particles, whereas today researchers can see the objects themselves—including surface and interfacing characteristics. This may be likened to Plato's cave, where one could only discern shadows cast by objects on the cavern walls, as opposed to direct observation of the objects proper. The latter scene is comparable to what occurs in nanoscale research, where practitioners now have access to information on the substances per se as given through form. In this science, intelligibility of observation and understanding are expressed as shape, which functions as the basic epistemological unit of nanoscale investigation.

In many disciplines in the recent history of the physical sciences, the parameter of force, such as Van der Waals forces, pervaded experiment and explanations. With the advent of nanoscale research, however, representation and analysis are frequently articulated as description where form lies at the heart of matter. This is not to deny the place of forces in NSR. But their significance is related to their impact on the determination of form and their transformations. Description signifies relating spatial connections between points that constitute an object or indicate spatial relations between entities. This is the basis of many of the results in nano. Here, the spatial scope of description is dimensionally small. Scientists attempt to disencumber the landscape. Description is confined to the acutely local and even specific. In one important respect, nanoscale investigations hark back to the later part of the nineteenth century, when descriptivism constituted a major analytic theme. The identity of nanoscale research thus entails looking backward as well as exploring "terra incognita."

For almost a century, epistemology of the microscopic world consisted of a probabilistic, stochastic vision. This meant that objects were described as huge populations of particles whose behavior was understood only collectively, and where the action of any individual elements was calculated in terms of chance. This way of comprehending is contrary to the major predicate of nanoscale research, where a deterministic perception underpins the endeavors of nanoscale scientists. This is one consequence of the aforementioned parameters of control, form, and descriptivism. Atom-by-atom and molecule-by-molecule manipulations are determinism in action. In the surface science of the 1930s, layers only a few atoms thick were successfully produced, but mainly analyzed and understood in terms of their collective properties or behavior. Nothing was known about local events. In nanoscale surface science, however, practitioners observe the relative positions of single atoms and molecules, can displace them and, by so doing, can determine the relations between form and positions and their effects on the local system.

A CHRONOLOGY

We here briefly present several episodes of the birth and development of nanoscience and technology. Although frequently evoked, these episodes are nevertheless elucidating. From Antiquity to the twentieth century, passing through the Roman Empire and the Middle Ages, there exist cultural objects that incorporate particles of nanometric scale. The first known example is an Assyrian sword blade in which the presence of natural nanoparticles reinforced the strength of the weapon. The Lycurgus Cup (fourth century BC) contains gold particles thanks to which the color of the cup changes (red when lit from behind, green when lit from in front). During the Middle Ages, nanoparticles were selectively incorporated to color stained-glass windows, for example, those of the Sainte Chapelle in Paris. Later on, the natural philosopher Michael Faraday (1791–1867) grasped the difference between optical effects of bulk gold and colloidal gold. He noted in 1847 that the presence of colloidal particles was connected with a change in the frequency of light as it shifted to red. This was the first studied observation of the effects of a nanodimensional substance.

At the turn of the twentieth century, Richard Zsigmondy (1865–1929) contributed significantly to the birth of nanoscale research: he invented the ultramicroscope (1890), which is capable of detecting nanometric objects, and he then used the device to explore the objects, learning that the dimensions of particles determine their physical properties and behaviors. Indeed Zsigmondy introduced the term "nanoscale." The observational capacity of the ultramicroscope opened many new fields of research; for example, it allowed Francis Perrin (1870–1942) and Albert Einstein (1879–1955) to experimentally validate Brownian motion. In the course of investigations conducted by Zsigmondy on the composition of colloids, a debate raged over whether they consist of crystals or particles. In order to settle the matter, Zsigmondy sought to produce a fine powder out of gold. After numerous attempts to reduce gold to the smallest possible division, he happened on Faraday's study in which the latter noted that, when exposed to light, gold in a finely divided form produces light of quite a different color—red in tint—than gold in its solid state or when it is composed of larger particles. Using a technique for obtaining very fine divisions of gold and employing his ultramicroscope, Zsigmondy studied the gold particles in a liquid. He determined that colloids consist of fine quantities and not crystals, which are much larger in size. Zsigmondy was thus able to see particles measuring only 10 nanometers (nm) in diameter that he could count for a unit of volume. He also gleaned some information about their form, noting that the form and size of matter strongly affect certain properties, notably optical properties. Hence gold in the form of nanoparticles produces red light, while larger quantities of gold yield the yellow light to which we are accustomed. More specifically, the scale and form of matter were interpreted as lying at the heart of physical characteristics and material behavior. In 1925, Zsigmondy was awarded the Nobel Prize in chemistry for his work on colloids and the invention of the ultramicroscope. During this period of nanoscale investigation, the distinction was established between the deterministic and the stochastic behavior of matter at this scale. As evoked above, this distinction has played a key role in much of the subsequent nanoscale research, as it suggests the possibility of controlling nanoscale objects.

Another event occurred in 1959 which, seen in retrospect, presaged the advent of nanoscale research. The physicist Richard Feynman (1918–88), Nobel Prize in physics (1965), gave an address before the American Physical Society at Caltech entitled "There is plenty of room at the bottom." In his talk, Feynman advanced three key ideas. Firstly, he claimed that it was possible to construct objects by assembling atoms one by one—the so-called bottom-up process. Secondly, engineers had long managed to build devices without possessing any thorough scientific understanding of their materials. This implies a refutation of the argument that, because we do not possess full knowledge of the world of atoms, we cannot build useful devices. Thirdly, implicit in Feynman's message is the idea that the world of molecules and atoms is deterministic and not unmanageably stochastic. Up to that point scientists had viewed the behavior of atoms as stochastic and as uniquely governed by the Heisenberg Principle. The suggestion that atoms and molecules might behave as deterministically governed particles ran contrary to the orthodoxy of the day. Finally, Feynman formulated what appeared to represent an impossible challenge: he proposed a prize of one thousand dollars for the first person who could build a machine in the space of a few microns, and a prize for the person who could encode the quantity of data corresponding to the *Encyclopaedia Britannica* within a space equivalent to the head of a pin. To the astonishment of all, the first challenge was successfully met within little more than a year, and the second, in 1985. Based on an arsenal of new materials, instruments, and learning, within fewer than thirty years this vision of controlling matter at the atomic and molecular levels became standard practice.

Another articulation of interest in nanodimensional objects was expressed through the term "nanotechnology," which was coined by the Japanese scientist, Professor Norio

Taniguchi, at the University of Tokyo in 1974. The term was introduced in the framework of investigation and fabrication associated with thin-film semiconductor deposition and ion-beam nanoparticle milling. The project here consisted of the fabrication of nanoscaled materials, which for the time required an unusual measure of control. The nanomaterials could serve both as vehicles of research and as components in manufacturing.

In an extensive series of articles and books, begun in 1981, the multidisciplinary MIT PhD engineering graduate Eric Drexler has set forth an unprecedented technological program intended to transform mankind's mode of material production. He argues the possibility of producing nanoscale machines capable of self-replication in the millions, which can, in a molecular dimension, manufacture more products than is presently conceivable. This approach would likewise lead to new kinds of products, and do so cheaply and without harming the environment or consuming today's vast quantities of energy. The program was first announced in his 1986 book *Engines of Creation: The coming era of nanotechnology*. In another book published in 1992, *Nano Systems: Molecular machinery, manufacture and computation*, the technically highly astute Drexler demonstrated the serious technical side of this program. He proposed precise designs of nanoscale motors, pumps, valves, and electronic molecular switches. The investigation and design is "theoretical" and did not involve experimentation and the construction of material artifacts. Throughout his studies at MIT and most of his subsequent research for NASA on space technology, Drexler's work was rooted mainly in computational techniques. He is a gifted computational engineer and over time his approach to nanoscale research and molecular manufacturing has become ever more based on computer simulation.

The endeavors of Drexler and his organizational empire, although not decisive, have proven highly important in the development of pro-nanoscale research policy and rich funding. The research, novel engineering concepts, "vision" of a very different material world grounded on nanotechnology-based molecular manufacturing, Drexler's organizational creations, his public militancy in favor of nanoscale-research government lobbying, and finally his testimony before many policy-related government committees have contributed importantly to the institutionalization and creative vigor of nanoscale research inside the USA. Throughout the 1990s and much of the following decades, he was a pivotal force in efforts to make molecular manufacturing an important theme in technological research, engineering, and enterprise.

Drexler's vision arose during the same period that the first nano instrumentation and nanomaterials appeared. Two key technical and scientific innovations that occurred during the 1980s underpinned the birth and rapid growth of nanoscale research. The first item, the scanning tunneling microscope, owes its unique place in the contemporary history of science to its capacity to observe single molecules and to control them with atomic precision. It is this capacity that has resituated determinism at the heart of epistemology and has given a new life to descriptivism. On another but related register, progress in existing technologies, such as lasers and ionization, has led to the synthesis of man-made novel materials. In 1985, the chemist Richard Smalley and his team synthesized the artificial nanometric substances that he termed "buckyballs." Their structure and remarkable properties sparked a flurry of research. Buckyballs initiated a new family of nanomatter, the fullerenes. Seven years later, the Japanese scientist Sumio Iijima synthesized the first carbon nanotubes. Such nano-objects have become a topic of massive academic research and are candidates for a long list of consumer products. Nano instrumentation has made it possible to study the new properties generated by these new materials, and the same devices have increasingly participated in the development of additional novel artificial substances. The number of practitioners engaged in nanoscale scientific research grew steadily through the 1980s. Beginning in the 1990s, nanoscale research expanded exponentially and today it continues to develop with an ever larger number of specialties.

In order to promote nanoscale research, the Foresight Institute was founded in 1986 by Drexler and Christine Peterson. This institute initiated the Feynman Nanotechnology Prize system in 1993 to recognize excellence in nano-investigation. Since 1997, one annual prize is awarded for experimentation and another for theory. A rapid glance at the prizes allows identification of key research domains. Four themes emerge: (1) Questions about the observation and the behavior of single molecules, and their connection with other molecules were particularly important during the first years of the Feynman awards. Such research was often associated with development of new or better systems of metrological instrumentation. Numerical experimentation in the form of simulation was also a central resource here, that opened the way to possibilities not immediately perceptible to metrology. (2) The entire history of the prize has been punctuated by research on the synthesis of new materials. Carbon nanotubes and nanowires appear frequently in prize attribution. (3) Nanobiology has received many prizes. DNA nanotechnology is important here. It is a material used for building wanted shapes. It is increasingly present in projects that develop bio-computational systems. Form is a significant component of nano, and nowhere more so than in nano work on proteins. (4) Finally, in line with Drexler's project of molecular manufacturing, numerous prizes have been awarded for the design and fabrication of nanomolecular machines. Some of these devices are single molecules, while others consist of an assembly. These are proto-devices intended to perform work, but whose powers are not established.

In 2000, the United States government formulated a science policy that richly financed nanoscale research. The announcement of the US Nanotechnology Initiative, at Caltech in January 2000 by President William Clinton, called for a significant reorientation in science and technology research and for the investment of billions of dollars by the federal government. Since the end of the Cold War, science and technology observers and experts, corporate leaders, and some in science-policy circles and academic research laboratories had addressed what they judged to be an increasing triple problem. This problem consisted of ever more pressing technological obstacles to the solution of material limits in the domain of microelectronics for purposes of computers and other electronic devices, and the associated endangerment of US corporate hegemony in these areas. Many held that, without radical innovation, Moore's Law, to which most clung, would be dashed. In this period of uncertainty, one observes the rapid and powerful entry of new players such as Japan and other competitors in these fields, with well-financed state research programs and corporate rationalization intended to enhance international competitiveness.

The sense of urgency to reform and rejuvenate application-driven programs in selected technology corporate fields was not shared by all at this time however, and in order to affect public policy it was necessary to identify key fields of possibly dangerous technical inadequacies, to propose promising paths of research, and to convince policy-makers, entrepreneurs, and relevant engineers and scientists to speak out and to act. For several years during the 1990s, the theme of giant magnetoresistance, discovered by Albert Fert and Peter Grünberg, for which they received the Nobel Prize in 2007, became one theme introduced as being capable of finding a solution to the current materials obstacle in computer memory, speed of data processing, and to other electronics bottlenecks. This approach temporarily found favor with policy-makers. However, in order to mobilize broader support for new science-and-technology policies and to obtain sufficiently high levels of investment, it was judged necessary to broaden the scope of appeal. The topic of giant magnetoresistance was thus somewhat eclipsed to be replaced by a far larger domain, one that could sometimes be presented as even an altogether fresh cognitive, material, consumer, and societal paradigm: namely the world of nanoscience, nanotechnology, nanomaterials, and consumer products.

In his Caltech speech, President Clinton announced that the US Nanotechnology Initiative would immediately receive a budget of two hundred million dollars—soon to rise to three hundred million; today it stands at almost one and a half billion dollars. Between 2005 and 2007, the US Department of Energy created multiple federal nanoscience- and technology-dedicated research institutes with annual guaranteed funding for ten years (20 million dollars), thus encouraging long-term programs. As the government of each nation carefully studies the research path taken by other nations, many European countries too developed nanoscience and technology (NST) initiatives, as did China. Today Chinese research institutes publish approximately 40% of the Thomson ISI Web of Science listed work on NST, the US publishing about 40% as well. Induced by easy funding, researchers have thus often hastened to modify their orientation in such a manner as to benefit from the fresh and entirely unanticipated flow of considerable resources.

In the absence of high levels of public funding, it is likely that NSR would not be enjoying the degree of expansion that it experiences today. Money speaks with a loud voice. Nevertheless, for many nanoscale research practitioners, funding represents a supplementary incentive, which adds to the quite independent attractions of a new cognitive frontier of artificial, made-to-order materials as objects of research and the call of a revolution in instrumentation, material, and epistemology.

Finally, it is only fitting to close our discussion with a word on nanotechnology products and hype. Countless articles have reported technological breakthroughs and products linked to nano: these include nanotechnology in water treatment, titanium nanoparticles in suncreams and nanomolecules as odor removers. In nanomedicine, for example, nano calcium phosphate is mentioned as an adjuvant for the delivery of bird-flu vaccines. There is also an immense amount of hype that clings to nano. The promise of nano superconducting materials for the transport of electrical energy has supplanted the earlier frustrated hopes for high-critical-temperature superconducting. Perhaps the best-known hype has been the promise of the use of fullerenes in the construction of geo-stationary space elevators. In both scenarios, nanoscience and technology occupy a central place in societal discourse and to some degree the realization of technical applications.

THE BOOK

In Chapter 1, "Mainstays of Nanoscale Research," the introduction of nanoscale investigation is discussed with reference to two key elements: (1) Development of two new experimental resources—scanning probe microscopy (the STM and the AFM) and numerical simulation; (2) synthesis of low-dimensional, nanostructured materials. Since the 1980s scanning probe microscopy has enabled scientists to observe microscopic objects of the nanoscale in much the same way that they had earlier observed objects in the macroscopic world. The frequent importance of form in classical science for the study of objects is echoed in nanoscale investigation, where the shape of molecules and molecular constellations lies at the heart of observation and intelligibility. Although in nanoscience much instrumentation is metrological, computational instrumentation equally participates in experiments. This coupling (metrology–simulation) illustrates one instance of what we call a combinatorial.

In parallel with scanning probe microscopy and simulation, new categories of nanostructured materials emerged during the late 1980s and 1990s: these include fullerenes and families of low-dimensional substances. They provide the matter investigated by practitioners and give rise to new categories of description and explanation. The metrological apparatus of nano is capable of molecular control as well as observation, and this ability further enhances the potential to synthesize new materials, which in turn spawn new questions and yield findings not present in the macroscopic world. This chapter describes a crucial novel technology for the synthesis of new materials, termed epitaxy. The introduction of this new cognitive/technical group, the "epitaxiors," who collaborate with metrological and computational experimenters significantly offers new possibilities specific to NSR. In the connection between metrological devices and materials, one again sees the centrality of combinatorials in nanoscale research.

Chapter 2, "Worlds of Nanophysics," opens with a description of the fascination with the possibilities and practices experienced by researchers during the birth and early development of nanoscale research in physics in the 1980s and 1990s. Scientists were stimulated by the power to observe the form and topology of single nano-objects and to control architecture on the molecular scale. As ever more nanomaterials were synthesized through control, the horizons of experimentation grew accordingly. It is this circulation between matter and measure that quickly came to constitute the motor of nano. Taken together, these factors precipitated a veritable "nano gold rush" from *c.*1985 to 2000.

In a process termed "research by design," we study the necessary relationships between experimenters and epitaxiors, where the direction of the historical link between matter and question is reversed. Important families of nanostructured matter are fabricated by the epitaxiors who painstakingly produce atomic-dimension objects layer by layer. In this nanoscale research, practitioners can order materials tailored to their cognitive curiosity, as opposed to being cognitively restricted by materials as they exist in nature.

Finally, this chapter explores the dealings between metrological and simulation experimenters. The emphasis by both instrument communities on molecular form and features comprises the shared focus and vocabulary that connect the two groups. Metrological experimentation often privileges sheer description of the research-object, whereas simulation work evokes the environment of objects and introduces complexity.

Chapter 3, "The Scale of Life?," describes the turn toward the possibility of observing the form of biological molecules and understanding the connection between form, molecular behavior, and function at the nanoscale. Nanoscale research indeed constitutes one important watershed in the evolution of molecular biology. The instrumentation of NSR now allows biologists to explore, indeed sometimes to see as visual representations, the very composition and configuration of DNA and proteins, and even to manipulate them. What we call the nanobiology "hexagon" in this chapter identifies six parameters that stand out when studying the research work in nanobiology. These parameters are (1) form/structure, (2) binding, (3) function, (4) three-dimensionality, (5) environment, (6) control. It is necessary to take into account all six of these to understand the realm of nanobiology where control constitutes a transverse framework.

In nanobiology, control is often expressed as molecular design, be it for DNA or for proteins. This is part of a new paradigm: to create shapes with nanometric biomolecules. Proteins can now be studied in terms of specific form, their dynamics, and with reference to their biological function. In this chapter we introduce the language and concepts of "lego" and "modularity," which are sometimes employed by nanoscientists and that prove particularly instructive in description and analysis in the domain. These elements underpin some work in evolutionary biology, where they serve to render compatible the intentionality of molecular design and chance. The 1965 Nobel Prize for medicine laureate, François Jacob, spoke of biological evolution in terms of tinkering as opposed to invention: tinkering is precisely what many nanobiologists do as they turn their gaze in the direction of molecular design. Design involves tinkering toward control. Nanobiology is of particular interest because it can reconcile these two superficially contradictory orientations.

Five characteristics of nanoscale research epistemology are set out in Chapter 4, "Epistemological Frames and Practices": form, image, descriptivism, local, and determinism. Scale and form constitute the primary entity of observation, description, and analysis. Three expressions of form are analyzed: morphology, occupancy of space and form, force and perturbation. The epistemological value of each category is discussed. In nanoscale research, images figure as importantly as today they do in culture at large. Images express aspects of spatial relations of objects, which, in practical terms, means form. They are the stuff of observation, description, and intelligibility. Their function is to provide information. The problematics of representation is not a consideration. Three categories of image production are presented in this chapter, followed by a detailed description of how image and form function in a specific nanoresearch project.

The complex relationship in nanoscale research between questions and materials is also an epistemological issue. In some instances, materials drive questions, while in others, questions induce the appropriate transformations of substances. The directionality of reasoning differs in the two scenarios. One can also observe an epistemology in which scientists transcend their habitual reflection on what materials imply as they consider the materiality of matter. As can be inferred from the above, the epistemology of nanoscale research revolves around considerations of determinism, descriptivism, and the local. Chapter 4 closes with a discussion of the place of theory and models in nanoscale research, and concludes that their contribution is quite circumscribed.

In Chapter 5, "The Role of Combinatorials in Structuring NSR Cognitive Trajectories," we document the centrality of combinatorials. The abundance of combinatorials in NSR magnifies the frequency of practitioners' circulation in the context of a phenomenon that we term "respiration." "Respiration" is a pause, a breath during which scientists evaluate, take stock of what new combinatorials are available for development of their present research project, or alternatively, that can be mobilized for initiation of a new project. This chapter explores scientists' recommitment to an existing research project or transfer to an alternative one. Researchers' reflections in terms of combinatorials thus determine "stop" or "go" decisions, and thereby fuel their cognitive trajectory. We describe operations of respiration in two different spheres. The "concentration" sphere is characterized by "epistemic expectations," which entails the funneling of all resources toward an understanding of the physical world. The second sphere, "extension," is grounded on what we call "function-horizons." The latter entails items like technical efficiency and reliability, industrial and state standards and norms, and markets. We discuss the alternative scenarios of intra-concentration respiration and of a concentration-extension respiration, in the light of the trajectory of two nanoresearchers. In one case, with reference to judgments about available combinatorials, the scientist remains inside the concentration sphere, but in the context of a sequence of respirations, he moves from project to project. In a second case, the practitioner similarly opts for circulation, but here moves from the concentration sphere to the extension sphere, where mobilization of function-horizons implies a radical change in combinatorials. In another respiration this scientist returns to the concentration sphere where epistemic expectations, combinatorials again prevail.

The final chapter of this book, Chapter 6, "Which Disciplinarity for Nanoscale Research?," seeks to locate nanoscale research in a competing landscape of disciplinary debate. Does it fit the standard disciplinary profile or instead correspond to emergent interdisciplinarity? We propose that there exists yet another alternative, which we term the "new disciplinarity." It is argued that this structure corresponds with the cognitive activity of most nanoscale practitioners.

The new disciplinarity entails six main features: referent, combinatorials, project, borderland, displacement, and temporality. The home discipline remains the principle referent of practitioners, but in nanoscale research, the horizons offered by combinatorials lead to projects with colleagues who are located in other domains. Multifarious projects result. Such projects are the crystallization of questions that cannot be resolved in a single domain. Scientists speak across the borderland whose structures allow them to remain inside their home discipline. The chapter closes with the presentation of the disciplinary positions of four nanoscale researchers. One of their trajectories fits the prescription of traditional disciplinarity, where borderland engagement is proscribed. The other three profiles represent alternative configurations of the new disciplinarity. In nanoscale research, we thus far observe no instance of interdisciplinary activity. Yet nanoscale research is a young science: time will tell.

RESEARCH METHODOLOGY AND TERRAIN

The originality of our research procedure resides in a combined scientific research articles/biographical methodology. This method revolves around scientists' published texts and practitioners' self-description of their intellectual evolution and collaborations. Our intention has been twofold: first, identification of the research questions, materials, instrumentation, and the epistemology that have underpinned a practitioner's scientific investigation. We view this quadrilateral as constituting a privileged geometry for understanding the contours, constraints, and opportunities of scientists' work in the nanoresearch metrics. This approach soon brought to light the centrality of a cluster of themes developed in our book, which include control, combinatorials, research-objects, form, respiration, and new disciplinarity. These themes are recurrent in publications and pervade discussions with dozens of scientists. Second, through a string of interviews, invaluable biographical information has allowed us to identify the complexity and range of interaction between groups of technological and cognitive specialists. Through selected biography, the structural components that underpin nanoscale research emerge in all their complexity and find expression in the contingencies of social and professional interactions and institutions that are here carefully set forth.

The descriptions, analyses, and conclusions of our research are based on interviews with 47 scientists engaged in nanoscale scientific research in multiple areas of physics and the life sciences, located in France and in the United States. Interviews revolved around scientists' instrumentation, research-objects, and questions. Exchanges ran for a minimum of two hours, and some scientists were solicited repeatedly.

In France, eight laboratories were contacted (30 practitioners): The Institut des Nanosciences de Paris (INSP) (22), the Laboratoire de Photonique et Nanostructures (LPN) (1), the Institut de Physique moléculaire de l'Université d'Orsay (2), the Centre de Génétique moléculaire de Saclay (2), the Institut de Biochimie et de Biophysique Moleculaire, Université d'Orsay (1), the Laboratoire de Physique Statistique, Ecole Normale Supérieure de Paris (1), Imagerie de l'Expression des Gènes (Commissariat à l'Energie Atomique) (1). In the United States, 16 Feynman Prize laureates were interviewed, plus one additional scientist: they are based at Northwestern University (3), University of California, Berkley (2), University of California, Los Angeles (2), California Institute of Technology (4), Temple University (1), Duke University (1), University of North Carolina (1), Georgia Institute of Technology (1), New York University (1), and one independent researcher.

1

Mainstays of Nanoscale Research

The now sizable and prospering domain of nanoscale research (NSR) may be regarded as an unintended consequence of two research projects. Each project yielded unintended results, and NSR is the child of the fusion of these twin unintentions. In this logic, NSR is itself a combinatorial, and combinatorials are one key signature of NSR. By "combinatorial" we refer to an association or interlocking of two or more components that give rise to a resulting novelty in the form of a synergy. The components that constitute combinatorials are multiple, including such elements as concepts, materials, instruments, methods, physical properties, themes, questions, etc. A combinatorial thus incorporates a diversity of items and can assume a variety of permutations, which in turn give rise to fresh combinatorials. While combinatorials probably figure importantly in much of contemporary science research, it is our contention that they are particularly significant in NSR, where they function as a motor.

This chapter documents the events linked to the birth of scanning probe microscopy, and in particular the scanning tunneling microscope (STM), and atomic force microscope (AFM), and it sets out to explain how such devices spawned important parts of nanoresearch. It will also describe the early developments of the synthesis of low-dimensional materials, which comprises the second component of nanoscience.

The first section of the chapter treats the circumstances that surrounded the invention of scanning probe microscopy. Here, the development of the scanning tunneling microscope rapidly gave rise to a concatenation of related apparatus that came to compose a family of metrological nano-dedicated instrumentation. This nano instrumentation was quickly joined by a second expanding species of instruments, namely, computational instruments in the guise of computer-driven numerical simulation. Scanning probe microscopy and simulation merged, forming a growing and powerful combinatorial for the investigation of nanoscale objects.

The second section of the chapter explores the extension of low-dimensional materials. Such materials were pursued because they exhibit physical properties absent from bulk materials. Low-dimensional materials are nanometer sized, and are thus welcome targets of study for nano instrumentation. Nanostructured substances began with fullerene buckyballs and carbon nanotubes, which were immediately joined by a flurry of other low-dimensional materials such as nanowells, nanowires, quantum dots, nanocavities, etc. Growing research on nano precipitated the extension and adoption of recent materials technologies such as molecular beam epitaxy and the birth of new materials techniques like self-assembling monolayers. The circumstances surrounding the birth of molecular beam epitaxy and self-assembling monolayers, and their place in NSR, are addressed in the third and closing passage of this chapter.

The relation between dedicated nano instrumentation and nanostructured materials has been further reinforced by virtue of the fact that STM and similar scanning probe microscopy apparatus can control the position of individual molecules and atoms, and can thereby contribute to projects for the generation of novel architectures entailed in new nanosubstances. In this complementary circulation and synergy between scanning probe microscopy and nanostructured materials, one observes a dynamic relation that initially spawned nanoscale research and continues to animate it.

1.1 INSTRUMENTATION DYNAMICS

1.1.1 Metrological instrumentation

What previous instruments and technological conditions accompanied the birth and rise of NSR? Why did NSR not emerge earlier given the fact that there have existed since the 1930s instruments such as the electron microscope, whose resolving power evolved toward one angström and even less—far smaller than what is required for nanometric investigations? The resolving power of the STM and related scanning probe devices, such as the atomic force microscope(AFM), is inferior to electron microscopy. This raises the crucial question of whether resolving power alone comprises the core of nanoscale research. Do other considerations not also intervene? But if so, which?

The ambitions, expectations, methodology, questions, concepts, and the very matter that comprise the multiple material building blocks of nanoscale scientific research are thus visibly linked to the two historically unanticipated recent occurrences: the startling development of an entirely novel kind of instrumentation and the sudden capacity to synthesize artificial low-dimensional substances. Novel nano instrumentation and nanomaterials frequently converge, and in doing so have come to develop combinatorials, which have multiplied the landscapes of knowledge growth and invention. The interplay between evolving instrumentation and materials has here spawned an unintended synergy, which has in turn led to more combinatorials. One may even discern in all of this the embryo of a nanoscale science "research system."

Much research carried out at the nanoscale focuses on molecules. Molecules comprise the basic units of physical and chemical organization and action, and it is the molecular environment that affects much photonic and electronic behavior. As will be seen below, the frame of observation of electron microscopy is ill adapted to molecular study. Moreover, electron microscopes are restricted in the range of substances they can explore. This emphasis on molecules does not diminish the importance of atoms in NSR. Scanning probe microscopy is capable of localizing single atoms, determining relative size and exploring defined clusters of atoms. Until the 1980s, science lacked a metrological apparatus adapted to the world of molecules as they function in a variety of environments and also able to precisely explore the kingdom of the atom. Unwittingly, it was the STM and associated instruments which aligned the capacity of detection in the nanometric window of observation for a simultaneous exploration of the form of individual molecules, their surface features, internal structures, and connections between molecules, and all of this in a variety of environments extending from ultra-vacuum, liquid, specialized gases and air, and in a thermal range stretching from near zero degrees Kelvin to several hundred degrees centigrade. Beyond this, they could displace single atoms and molecules such that they may constitute novel architectural arrangements, and through this comprise new artificial material compositions. The material compositions were in turn available for exploration by the same STM instrument from which they sprang. The radical nature of this nano instrumentation is that it is both "passive" in its observation function and "active" in its capacity to manipulate atoms. These qualities of circularity and synergy have emerged strongly in NSR, and the relation has substantially fueled the field.

NSR metrological experimentation is driven largely by two categories of devices, the scanning tunneling microscope and the atomic force microscope. Taken together, these and related instruments constitute a family of apparatus called scanning probe microscopy.[1] The parameters of descriptivism, control, and the detection and observation of single objects such as molecules and atoms are the purview of these instruments. In NSR, instrument control is inextricably connected to construction. This link between instruments and construction finds expression in the synthesis of novel nanometric materials, which are the subject of the second section of this chapter. NSR is a field where matter and instruments are inseparably combined and may even be regarded as constituting a whole.

The emblematic STM was developed in 1981, and the potential and success of the metrology are testified to by the Nobel Prize awarded to its inventors in 1986—a mere five years after the birth of the apparatus.[2] Research for the STM was carried out inside the Zurich laboratory of IBM by two employee physicists: Gerd Binnig and Heinrich Rohrer. What became the STM was a byproduct of a company project to revolutionize computer technology through creation of a new species of superconducting semiconductors that would out-perform extant transistors, both in terms of speed and thermal dissipation characteristics. In the late 1970s IBM, like other computer firms, anticipated blockage in computational performance due to current physical properties of extant transistor technology, whose speed and quantitative capacity were nearing a threshold. The company's research project was intended to invent superconducting high-performance transistors.

[1] L. Aigouy, Y. De Wilde, C. Frétigny (2007) *Les Nouvelles Microscopies. A la découverte du nanomonde*. Paris: Belin; C.C.M. Mody (2011) *Instrumental Community. Probe microscopy and the path to nanotechnology*. Cambridge, MA: The MIT Press.

[2] C.C.M. Mody (2011) Instrumental Community. G. Binnig, H. Rohre (1993) Scanning Tunneling Microscopy – From birth to adolescence, Nobel Prize Lecture 8 December 1986. In: Physics 1981–1990, Editor-in-Charge Tore Frängsmyr, Editor Gösta Ekspang, Singapore: World Scientific Publishing Co.

A superconducting state eliminated issues of heating and would multiply capacity for previously unattained computational potential. This technology would be based on the Josephson effect, in which electrons selectively pass through the surface boundary layer of a semiconductor. The IBM program designed and sought to construct an electronic Josephson junction device, consisting of several layers of superconducting semiconductor materials separated by a very thin insulating band. Unfortunately, the insulating layer was plagued by the irregular presence of pinholes, which permitted passage of unwanted electrons between the semiconducting regions.[3] Detection of these regions meant the identification of nanoscale structures. For an STM, good resolution is now considered to be 0.1 nanometer (nm) lateral resolution and 0.01 nm depth resolution. In what is referred to as its "scanning mode," the STM can now locate individual atoms and identify the position and the surface characteristics and structures of individual molecules. After surmounting considerable obstacles, the instrument proved capable of identifying and localizing the target pinholes in the insulating band.

It is to be noted that, throughout most of the research that led to the STM and to its construction and trial testing, the company came close to turning its back on STM work.[4] In spite of this internal company indifference and even reluctance toward the device, its inventors nevertheless sought to publish a scientific journal article on the detection characteristics and research possibilities of the instrument, yet to no avail. The manuscript was rejected. But in spite of this, through direct exchange with small groups of scattered scientists, Binnig and Rohrer slowly and painfully began to convince a few researchers, often working in the domain of surface science, of their instrument's performance and possibilities. These efforts were entirely outside the purview of IBM and established channels of institutionalized science; and they touched few practitioners. The exercise proved complicated because the IBM machine was initially the only existing instrument. As indicated by Cyrus Mody, although technical information on components and construction rapidly became publicly available, individuals interested in trying the STM had to improvise as best they could with the components at hand.[5] Nevertheless, by 1983–4 numerous practitioners had managed to build their own devices and quickly discerned the potential horizons of this revolutionary apparatus. Following on subsequent rapid success in surface physics (see Chapter 2), the instrument was soon perceived to offer experimental hope in crystallography, magnetism, low-dimensional materials, domains of chemistry, and of utmost significance, a few years later, even molecular biology (see Chapter 3).[6]

[3] C.C.M. Mody (2011) Instrumental Community. G. Granek, G. Hon (2008) Searching for asses, finding a kingdom: the story of the invention of the Scanning Tunneling Microscope (STM), *Annals of Science*, 65 (1): 101–125.

[4] IBM ultimately abandoned the superconducting semiconducting project. C.C.M. Mody (2011) Instrumental Community.

[5] C.C.M. Mody (2006) Corporations, universities, and instrumental communities: commercializing probe microscopy, 1981–1996, *Technology and Culture*, 47(1): 56–80; C. Joachim, L. Plévert (2008) *Nanosciences. La révolution invisible*. Paris: Editions du Seuil.

[6] C.C.M. Mody, M. Lynch (2010) Test objects and other epistemic things: a history of a nanoscale object, *British Journal for the History of Science*, 43(3): 423–458.

IBM had intended the STM to be purely a detection device for the localization of pinhole insulator defects. In the event, the potential of the instrument lay elsewhere. Its outstanding strengths instead resided in alternative capacities. The STM is capable of observing the form of single molecules, their surface characteristics, and sometimes even their internal structure, and of determining the geometry of interfacing between molecules. This power of topological and broader morphological discrimination for individual molecules and collectives enabled the instrument to function as a key to the world of nanodimensionality. This emphasis on the centrality of the molecule is not intended to mitigate the STM's ability to explore the presence and location of atoms, where it provides rich information on relative position and dimension. The device offers a view of the atomic landscape essential to the solution of important physical questions.

Here one clearly sees that the achievements and potential of the STM lay far afield from the intentions and expectations of IBM for the apparatus to operate as a defect detection machine. Cyrus Mody has clearly documented that the mission of the STM in science is thus an unintended outcome of engineering and metrology efforts.[7] The contributions to nanoscience and to knowledge at large are accidental, but nevertheless accidents along a line whose orientation and content exhibit directionality.

The STM afforded a second key input to nanoscience. In its "point contact mode," the device can precisely manipulate atoms and molecules, and through such control, it can construct novel molecular objects. Point contact consists of the selective displacement of an atom or other microscopic entity due to the action (shoving or dragging) of the STM tip (to be described below). As will be seen throughout this book, control at the molecular level is central to NSR and is principally made possible by the STM, which can displace selected entities precisely from one position to another. Such manipulations permit the construction of predesigned architectural nanostructures and shapes that will subsequently be explored during research on unanticipated physical properties.[8]

The STM has often been heralded as the device that has made it possible to "see" the atom. History proves more nuanced and complex, however. By the 1980s when the STM began to observe atoms, it was the case that other instruments had already achieved this objective. The field ion microscope had been invented in 1951 by Erwin Wilhem Mueller (1911–77), which in 1955 spotted for the first time in history the presence of a single atom.[9] By the 1970s the transmission electron microscope (TEM) and scanning electron microscope (SEM) possessed the same capacity for restricted categories of substances,

[7] C.C.M. Mody (2011) Instrumental Community.

[8] Our discussion of "control" in nanoscale science is not meant to evoke the fruitful concept of "thing knowledge" proposed by Davis Baird. In nanoscience, manipulation does not operate as cognition. It addresses an alternative set of expectations and functions. It is principally a means to an end, constituting a powerful capability in the arsenal of materials synthesis and the fabrication of scientific research objects. D. Baird (2004) Thing Knowledge. *A philosophy of scientific instruments*. Berkeley: University of California Press.

[9] <http://en.wikipedia.org/wiki/Field_ion_microscope> (accessed 17 May 2013).

subsequent to complicated sample preparation. Some scholars judge that the metrological potential of TEM and SEM was indeed similar to the STM during the 1980s, and that they could thus have been put to far better use in research.[10] In appreciating the place of scanning probe microscopy in nanoresearch and the recent history of science, it is essential to understand that, in the case of the TEM of the 1950s and the SEM of the 1960s and 1970s, atoms were principally viewed as spots; as points.[11] Their presence could be detected, but their properties could not be explored in minute detail. The originality of the STM and other instruments in the family of scanning probe microscopy lay principally in the breadth of information that could be gleaned, in the range of substances that could be explored, and in the multiple environments that could effectively be dealt with. This latter introduced a huge universe of properties and effects for study, unavailable to all former categories of apparatus. "Seeing" the atom is important, but what exactly is being seen and to what end? Finally, pre-scanning probe instruments may be described as "passive" devices because, like the vast majority of metrological apparatuses, they simply observe objects. That is to say they cannot be used to manipulate them.

By contrast, scanning probe microscopy is a "dynamic" instrument. It is in part a control device: probes can be employed to manipulate the position of individual molecules and atoms, and to construct new arrangements in order to generate or measure new physical effects. It is also observationally dynamic: probes are moved about in order to adjust observational perspective in terms of changing experimental conditions or demands. Finally, on an epistemological level, although probabilistic quantum analysis and descriptions of atomic and electronic behavior have long colored perceptions and work on the microscopic world,[12] hands-on practical experimental practice with the STM of seeing, understanding, working, and building quickly engendered a deterministic perception of the nano world.

[10] D. Baird, A. Shew (2004) Probing the history of scanning tunneling microscopy. In: D. Baird, A. Nordmann, J. Schummer (eds.) *Discovering the Nanoscale*. Amsterdam: IOS Press, pp.14–156.

[11] Throughout the 1960s and 1970s much of the experimental research in the domain of surface science was conducted using low energy electron diffraction (LEED) and photo-electron spectroscopy (PES). In the case of LEED, particles were bounced off the surface of crystals, and the measured angle of diffraction generated information on the relative position of atoms in the substance. PES measured the energy distribution yielded by excited surface atoms, thus promoting knowledge on the atomic structure of the material. Both of these key devices of the time indicated regions of regularity. They operated in reciprocal space where intervals between entities are indicated versus information on the shape, size, texture, etc. of the target object. They thereby provided "indirect" information instead of a "direct image."; <http://fr.wikipedia.org/wiki/Low-energy_electron_diffraction> (accessed 17 May 2013).

[12] M.J. Nye (1972); M.J. Nye (1994) *From Chemical Philosophy to Theoretical Chemistry: Dynamics of matter and dynamics of disciplines, 1800–1950*. Berkley, CA: University of California Press; A.D. Wilson (1991) Mental representations and scientific knowledge: Boltzmann's Bild theory of knowledge in historical context, *Physis*, 28: 770–795; M. Jammer (1974) *The Philosophy of Quantum Mechanics: The interpretations of quantum mechanics in historical perspective.* New York, NY: Wiley-Interscience; id. (1989) *The Conceptual Development of Quantum Mechanics*. New York, NY: McGraw-Hill (2nd ed.; 1st ed. 1966); C. Joachim, L. Plévert (2008) Nanosciences.

Although the novel and perhaps radical historical capacities of scanning probe microscopy as incarnated in the STM and AFM are today readily perceived, what are these devices built of and how do they operate? An STM consists of six major features:

(1) The quantum-mechanical electron tunneling effect (the Josephson effect),[13] which lies at the heart of the STM's performance, can be described as a cloud of electrons that escapes a material's surface boundary layer. A few electrons tunnel through this surface layer of the object. In the case of the STM, tunneling electrons are emitted by both the target research sample and by the instrument's tip. Tunneling data, in the form of current and voltage, are used to identify the position of individual atoms and molecules, and to describe the topography of surfaces of the latter.[14] This instrument does not "see" atoms or molecules as do conventional optical microscopes, but rather "senses" particles by registering minute changes in electric current.

(2) The tip depicted in the diagram (Figure 1.1) is a reading device, which detects the position of atoms and molecules. It consists of a single atom and it is extremely fragile and subject to damage.

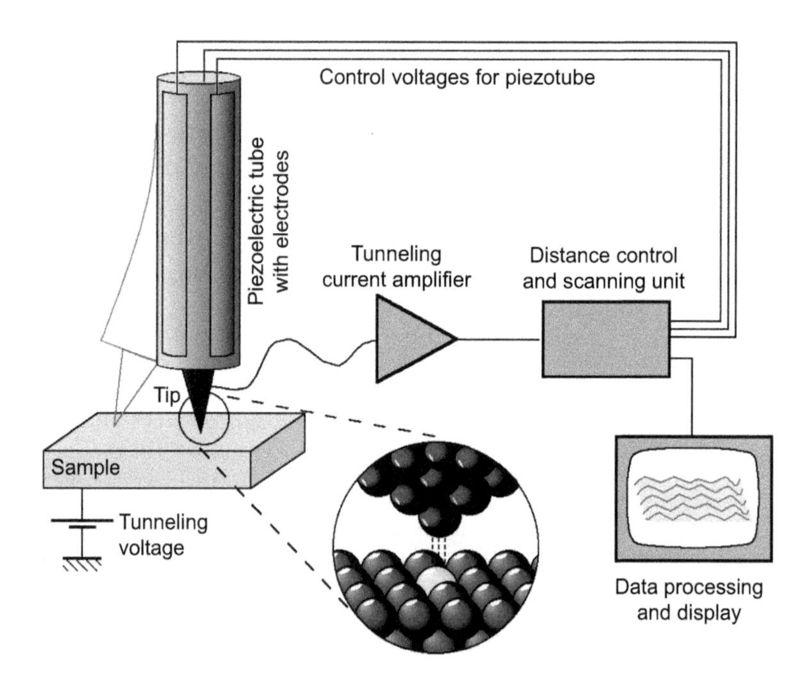

Figure 1.1 Scanning tunneling microscope schematic. Image by Michael Schmid, TU Wien, courtesy of Wikimedia Commons.

[13] Brian David Josephson received the Nobel Prize in 1973.

[14] D. Rothbart, J. Schreifels (2006) Visualizing instrumental techniques of surface chemistry. In: D. Baird, L.C. McIntyre, E.R. Scerri (eds.) *Philosophy of Chemistry: Synthesis of a new discipline*. Dordrecht: Springer, pp. 309–324.

(3) The position of the tip is computer controlled along the x, y, and z axes with reference to the sample surface.
(4) The STM explores objects through multiple scans in a rafter pattern, and this is also controlled by computer.
(5) These data are next processed, organized, and structured by a series of specialized algorithms.
(6) Computer graphics transforms these data in order to depict them as images that represent the position, relations, and topology of the object under study.[15] The complexity of the STM is suggested in Figure 1.2.

The device gives the object in terms of location and form on a visual display unit such as a computer screen. The organization of information as three-dimensional morphologies introduces a sense of proximity and solid reality that differs from numerical and graphical information. It permits practitioners to "see" details of objects and relations between them, and to swiftly shift from analysis of a part to reflection on the whole.

Figure 1.2 Photograph of an STM. Image reproduced with permission of Argonne National Laboratory.

[15] In early models of the STM, information on the motion of the tip was conveyed by a chart-recorder and presented graphically.

Figure 1.3 Eigler's spelling of IBM with xenon atoms. Image originally created by IBM Corporation.

This once again affords a new perspective on the micro world that invites fresh wonder and earlier unheard questions and speculation. As shown in Chapters 2, 3, and 4, visual images constitute an important platform for reasoning in nanoscale research.

The information generated by the STM and related devices has made it possible for scientists to reason about and to grasp many aspects of the nanoscopic world, in ways that can be perceived and analyzed in the grammar of the macroscopic world. The introduction of the STM into contemporary science has in this way promoted a kind of epistemological revolution.

This epistemological revolution has had a psychological impact on scientists. In 1989, the physicist Donald Eigler, employed at IBM, published an article in *Nature*, featuring an image of the letters I.B.M., which were obtained using an STM tip to precisely position 35 atoms of Xenon to form the three letters.[16] Many scientists interviewed in the course of our research project testified to their amazement at "seeing" images of single atoms that had been individually manipulated to form a designated stable pattern (see Chapter 2). The precision of atomic control obtained by Eigler is readily observable in Figure 1.3. What emerges here as a dramatic and profound impact upon witnessing these three atomically constituted letters of the alphabet may in part be rooted in the centrality of inscription and writing in the history of our culture.[17] This episode may also be seen as announcing the new place for visual images in the constitution and communication of contemporary microscale research.

In concrete terms, what kind of novel achievements are being reaped through STMs?[18] In past decades, research on crystal structure often dealt with measurement of the alignment of crystals or the angles between them. Using the STM, crystal study now focuses on the precise location and arrangement of individual atoms and molecules, whether they are jumbled or strung out, the existence of flatlands, mountains or plateaus, polarization, etc. The STM and its cousin, the scanning tunneling spectroscope (STS) have been

[16] D.M. Eigler, E.K. Schweizer (1990) Positioning single atoms with a scanning tunneling microscope, *Nature*, 344 (6266): 524–526. He received the Kavli Prize in 2010 for his breakthroughs.

[17] A. Marcovich, T. Shinn (2011a) Instrument research, tools and the knowledge enterprise –1999/2009. Birth and development of dip-pen nanolithography, *Science, Technology, and Human Values*, 36(6): 864–896.

[18] C. Joachim, L. Plévert (2008) Nanosciences.

effectively mobilized to describe and study properties of carbon nanotubes.[19] This substance exhibits a range of architectures precisely indicated by the STM, which includes information about thickness and wrapping architecture. Architecture determines whether the material is a conductor or a semiconductor, and it affects its mechanical properties. Together the STM and STS map the electrical characteristics of carbon nanotubes and of other low-dimensional material (see the second section of this chapter).[20] Finally, thanks to the STM, and even more frequently its sister device, the atomic force microscope (to be discussed below), nanobiologists today explore the morphology of single proteins, where they can discern the shapes that prohibit or allow binding and study the path along which binding occurs. Similarly, the STM and related devices can visualize DNA collectives and structures, sometimes in a particular bio environment (see Chapter 3).

The second metrological instrument emblematic of NSR is the atomic force microscope (AFM). The atomic force microscope was invented in 1985 at Stanford University by Gerd Binnig (co-inventor of the STM), Calvin Quate (Stanford University and associated with the STM's early developments), and Christoph Gerber (IBM, Zurich). It was commercialized in 1989, some eight years after the introduction of the STM.[21] As shown in Figure 1.4, an elementary presentation of the atomic force microscope draws attention to five main components:

(1) A cantilever constitutes the heart of the device which is fundamentally a mechanical deflection/vibrational apparatus.

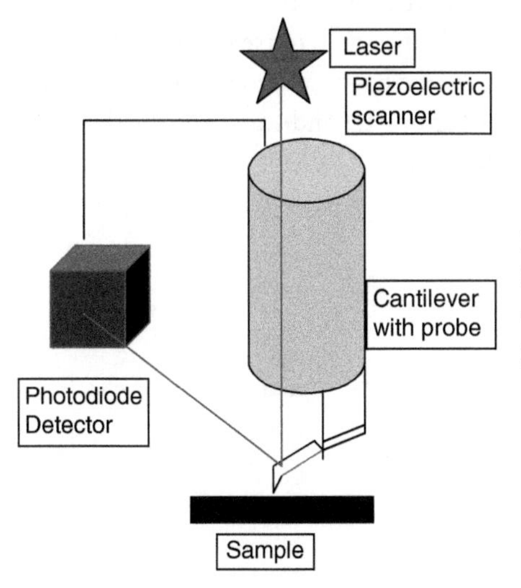

Figure 1.4 Schematic depiction of an AFM instrument showing "beam bounce" method of detection using a laser and position-sensitive photodiode detector.

[19] Scanning tunneling spectroscopy is closely related to its cousin STM. The STS measures energy distribution, as do spectroscopic devices in general, as opposed to measuring the position of objects and spatial relations between them. While an important nano-instrument, it is far less commonly used than the STM and other metrological devices.

[20] J.W.G. Wildoer, L.C. Venema, A.G. Rinzler, R.E. Smalley, C. Dekker (1998) Electronic structure of atomically resolved carbon nanotubes, *Nature*, 391(6662): 59–62.

[21] C. Mody (2011) Instrumental Community: Probe Microscopy and the Path to Nanotechnology.

(2) The cantilever is equipped with a tip which registers properties of the sample surface by interacting with it, and which induces cantilever deflection.

(3) A laser beam signals the mechanical/vibrational motion of the cantilever. This motion is the vehicle that transmits information from the sample concerning topography and force fields to data-collection and processing circuits. The word "force fields" is of foremost significance. The instrument's capacity to detect and measure multiple force fields extends its range of relevance.

(4) The laser beam is reflected from the cantilever to a photodiode detector. The diode data drive the control and feedback circuit and provide computers with inputs.

(5) Computers intervene in the piezoelectric and feedback systems, photodiode, and for final data processing and generation of visual images.

The device offers several detection modes: (1) the "contact" mode, where the tip operates at a distance between 0 and 0.5 nm from the sample, (2) the intermittent "contact" (tapping mode), between 0.5 and 2 nm, and (3) the non-contact mode between 0.1 and 10 nm. Mode selection depends on the forces designated for measurement.

Although the STM is incontestably the instrument most frequently associated with the rise and spread of NSR, it is not the device most often used in nanoscale research; this is the atomic force microscope. The number of articles based on the STM climbs steadily between 1982 and 1998, after which the curve is more or less stable. Between 1990 and 1994, use of STM quadrupled, climbing from about 400 to almost 1600 annual articles. As will be shown in the next chapter, many practitioners rushed to the new device and used it to explore almost any material that came to hand. Over the course of its life, the STM has figured in 40,000 articles as against the AFM, which has appeared in over 88,000 publications—double the number of the STM! (See Figures 1.5 and 1.6.) By 2000, the AFM was already enjoying a comfortable lead over the STM.

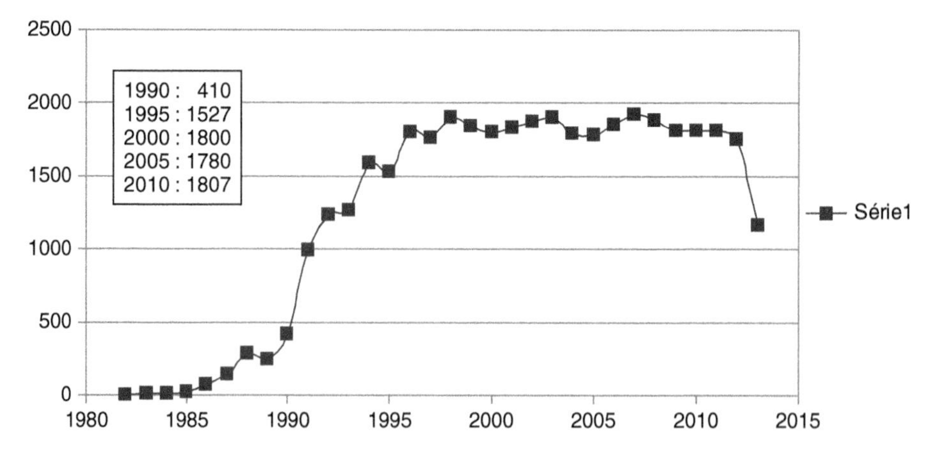

Figure 1.5 Scanning tunneling microscope: Number of publications: Topic: scanning tunneling microscope: 40,240 items (Thomson ISI Web of Science: 26 September 2013).

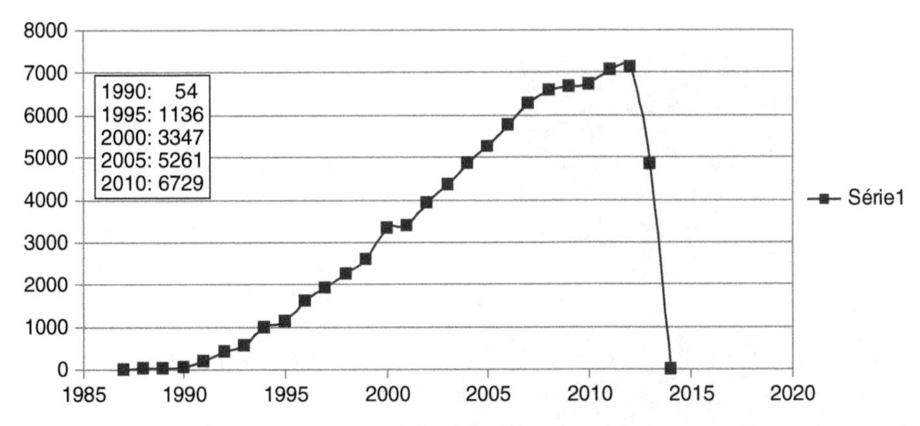

Figure 1.6 Atomic force microscope: Number of publications: Topic: atomic force microscope*: 88,000 items (Thomson ISI Web of Science: 26 September 2013).

What exactly accounts for the prevalence of the AFM in NSR? Is it an issue of resolution, stability, or precision? Is it the families of materials it can explore? Does the need to prepare samples for analysis thus introducing experimental artifacts figure in the selection? Are issues of highly specialized training to use the device, cost, etc., relevant? In other words, what are the specific contributions to research and learning unavailable to other metrological instruments? Part of the answer to this important question is connected with the AFM's technical characteristics and associated operating modes.

Unlike the STM, which often entails training and considerable experience for effective utilization, the AFM does not demand an instrument specialist. It is frequently a tabletop instrument, sufficiently stabilized and standardized that advanced knowledge, technological expertise, and long experience are not required to use it. Figure 1.7 illustrates the compactness and portability of the device. This opens new functionalities and horizons.[22] The ability to read a range of forces proves decisive. Depending on experimental needs, the forces measured by the AFM include mechanical contact force, Van der Waals forces, capillary forces, chemical bonding, electrostatic forces, magnetic forces, etc. Along with force, additional quantities may be measured simultaneously through the use of specialized types of probe.

Most importantly for the success of the AFM, multiple and diverse families of materials and alternative physical properties are identified and characterized through emission of a range of force fields. While the STM is principally limited to metrology of semiconductors and conductors in general, the AFM can deal with both semiconductors and insulator materials.

In addition, complicated and sometimes damaging preparation of samples is unnecessary. This is particularly decisive in the case of biological research, where molecules such as proteins or DNA are destroyed by conditioning such as desiccation, vacuum,

[22] A. Marcovich, R. Shinn (2011a) Instrument research, tools and the knowledge enterprise.

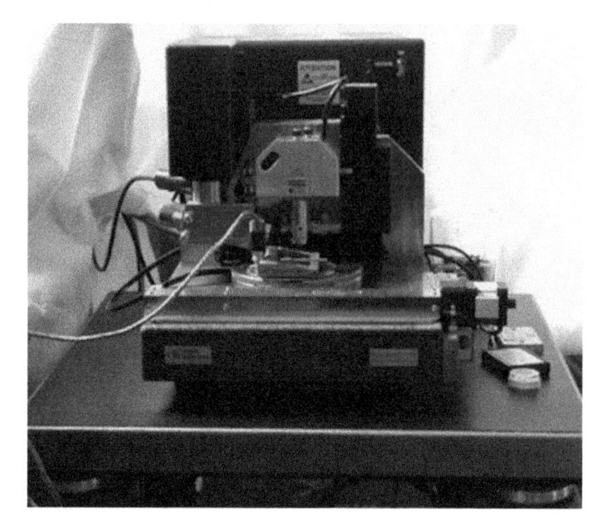

Figure 1.7 Photograph of an AFM. This is a Veeco Dimension 3100 AFM/MFM instrument using specially constructed magnetizing stage at the Kelvin Nanocharacterisation Centre, University of Glasgow, UK.

ultra-cold, and metallic coating. Moreover, this device is capable of observing in vivo systems (see Chapter 3).

It is safe to say that NSR has become a far-flung empire, in part thanks to the many families of materials that the AFM can explore, to the physical parameters that it detects, and to the range of measurements (mechanical, magnetic, electronic) that it allows (for example see Chapter 5).

Although scanning probe microscopy, and in particular the AFM, is emblematic of NSR metrological instrumentation, a close inspection of the devices employed in nano experimentation and which today punctuate much publication reveals that electron microscopy also abounds. This fact does not contradict our above argument that observational limitations of electron microscopy are such that molecular study is marginalized for reasons of excessive resolving power. The nanoscale is indeed principally molecular in dimension, and this dimension belongs to the AFM and STM. In spite of this, electron microscopy definitely has its place in nanoscience, a very specific place that is often complementary to scanning probe microscopy.

Precisely how does the electron microscope contribute to nanoresearch, and why is it not commonly viewed as pivotal and on a par with the STM? Indeed, a careful look at the record clearly shows that TEM and SEM participate significantly in certain key aspects of the field of nano. It is notably important for the verification of synthesized nanomaterials. This instrumentation detects atomic positioning and linkage, thus complementing STM information. Its great strength lies in detection of defects in materials that may arise from atomic misalignment. So the value of TEM and SEM resides in monitoring functions. Electron microscopy is often associated more with industrial activities than with academic efforts.

The electron microscope was developed by the German physicist Ernst Ruska (1906–88) in 1933. He won the Nobel Prize, along with Binnig and Rohrer, in 1986. This is symbolic of the connection and combinatorial between scanning probe microscopy and electron microscopy. The first commercial electron microscope was built by the Siemens Company in 1939. Today electron microscopes are capable of two million power magnification. Its first applications lay in the area of biology and medicine, and this device proved central to the progress of biological research at the Rockefeller Foundation during the 1940s and following. In the 1950s and 1960s, relatively large numbers of highly standardized electron microscopes were manufactured, commercialized, and put into service. Today's electron microscopes are capable of 0.5 angström resolution.

An electron microscope consists of a high energy electrostatic and magnetic "condenser lens" that controls an electron beam (electron gun) which focuses on a specimen and thereby produces an image on a fluorescent screen and camera. Electron microscopy is founded on the use of the electron's wavelength, which is a thousand times shorter than that of light, and this explains why electron microscopy has a much higher resolving power than standard optical microscopes. It involves the passage of focused electrons through specific regions of a sample and blockage in other regions. Electron study is restricted to ultra-thin samples. In many instances obtaining a thin sample and opacification destroys the very structures or features intended for study. Imaging is thus carried out on crystals or metal-related substances. Investigation of biological material is complex, and in situ study is currently impossible with a TEM or SEM. With the advent of the computer, a variant of the electron microscope was spawned, namely the scanning electron microscope. This device generates images based on multiple-rafter scans where the final representation of the object is a construct based on an assembly of the many images. Scan control and image assembly entail the association of a computer.

The immense success of the two categories of electron microscopy is in part attributable to the high quality of images. When compared with scanning probe microscopy, their representations render a great amount of detail. Another strength lies in the fact that electron microscopy constitutes a long-standing community marked by the existence of many devices and well-trained operators. This contrasts with the thirty years' history of scanning probe microscopy and nanoscale research. For a thorough understanding of the emergence and evolution of NSR, it is essential to grasp that TEM and SEM are significant because of their capacity to monitor the quality of synthesized nanostructured materials.

There exists a systemic difference between the logic of electron microscopy and NSR. Electron microscopy belongs to "big science," as opposed to most laboratory work in NSR which can be accommodated on a desktop or located in a room, albeit a sizable one. Electron microscopy is extremely expensive. It entails cumbersome devices that require a vast and complex division of labor to operate the machines. These instruments also necessitate fastidious preparation of samples. All of which implies an unwieldy organizational framework.

Examination of the most-cited articles listed in the ISI Web of Science associated with TEM and SEM clearly shows that much work is oriented to the characterization of materials, to engineering-related questions, and to measurements linked to industrial production and quality control. Far fewer publications deal with the physics of materials in terms of properties and dynamics of phenomena. The technical and cognitive landscape of electron microscopy and the metrology of scanning probe microscopy thus differ significantly. As indicated above, in NSR experimental investigations, scanning probe microscopy offers many distinct advantages. It allows the manipulation and control of objects that can render a kind of insight not available to electron microscopic investigation, which is static. Because the probe can be moved about, scientists can also spontaneously focus on sites that call for closer inspection and then move to investigate other sites during a process of landscaping and comparison.

To summarize, scanning probe microscopy has made it possible to observe size, form, surface characteristics, and spatial relations of single molecules, and relative positions of atoms, and to identify forces functioning at the nanoscale. Nanoscale objects are seen to behave deterministically, and can thus be manipulated accordingly. Taken together, the powers to control and to observe a minuscule detail on the nanoscale now allow researchers to deal with nano objects in some respects along the same lines that have long been the case with bulk materials. To the extent that representations, forces, and capacity for control associated with bulk matter are now the coinage of research and engineering at the nanoscale, it is perhaps justified to speak with confidence of a nano revolution. The second indisputable ingredient of this revolution is the synthesis of entirely new categories of materials (for example low-dimensional materials having zero, one, or two dimensions) that do not exist as bulk materials. These novel nanoscale-structured objects often exhibit amazing properties linked to their dimensions—properties that had never before been imagined. These materials and properties will be discussed below in Section 1.2.

1.1.2 Computational instrumentation and numerical simulation

Numerical simulation constitutes the second family of instrumentation (computational instrumentation) in experimental nanoscale research. Although simulation is definitely a stand-alone device, it nevertheless often operates in conjunction with metrological apparatus. The result of such simulation / metrology experimentation is a combinatorial of huge potential and power. This combinatorial will frequently be described in later chapters.

Simulation entails mathematical models that attempt to develop "discretization" solutions to a problem, and thereby enable the description and prediction of the behavior of the system under examination from a set of parameters and initial conditions.[23]

[23] P. Humphrey (1995) Computational science and scientific method, *Minds and Machines*, 5: 499–512; F. Rohrich (1990) Computer simulation in the physical science, *PSA: Proceedings of the Biennial Meeting of the Philosophy of Science Association*: 507–518; P. Humphrey (2004) *Extending ourselves*. N.Y.: Oxford University Press; Computational Science, Empiricism, and Scientific Method. J. Lenhard, (2004) Nanoscience and the janus-faced character of simulations. In: D. Baird, A. Nordmann, J. Schummer (eds.) (2004) *Discovering the nanoscale*. Landsdale, PA: IOS Press Inc., pp. 93–100. P. Hymphrey (2009) The philosophical novelty of computer simulation methods, *Synthese*, 169(3): 615–626; T. Grüne-Yanoff and P. Weirich (2010) The philosophy and epistemology of simulation: a review. *Simulation & Gaming*, 41 (1): 20–50; J. Lenhard (2007). Computer Simulation: The cooperation between experimenting and modeling, *Philosophy of Science*, 74(2): 176–194.

Parameters are given numerical values, and their relations and hierarchy are assigned. Physical and even computational simulation originated long before the advent of nanoscale research,[24] but we shall see that it is particularly well adapted to the latter, and that simulation largely contributes to the claim that NSR is a synergistic science.

Simulation draws together multiple components. It aligns contributions from electronic numerical calculation, development of algorithms and special models, and the emergence of computer-based imagery, often termed "computer graphics." Simulation extends well beyond nanoscale research.[25] It can be arranged into three principal compartments: entertainment (films and games); structural and product design–process engineering and testing; and finally simulation in scientific research. This last field came to fruition relatively later than the other two fields. In spite of its recent introduction, simulation in science has become a large and important area of endeavor and community. Its status in science stands between mathematics and theory on the one hand and physical experiment on the other. Exactly how significant simulation is as a research instrument in nanoscale research can be assessed by comparing it with other devices. Since the early 1980s the number of nano-related articles linked to simulation stands at about 50,000 publications, compared with 40,000 for the STM and 88,000 for the AFM.

In general, how does simulation operate in nanoscale research? Before addressing this important question, a brief pause is required in our narration in order to sketch how simulation functions in general and through this its relevance to scientific research. Since the late 1920s, equations have existed in quantum mechanics associated with wave functions of atoms and molecules that can, under certain conditions, provide approximate information about the structure and properties of matter. However, these equations are extremely complex so that the calculations cannot be carried out by scientists working with pen and paper or with elementary calculating machines.

The solution to such equations depends on the use of powerful and fast electronic computers and the development of appropriate algorithms and models. Computational simulation in scientific research entails the development of an algorithm compatible with the equation that allows solution for specific aspects of the particular material under investigation and with reference to designated physical properties. Simulation scientists have devised a model that expresses relevant physical conditions such as total atomic or molecular energy, thermodynamic conditions, temporality in some instances, elements of the environment, etc. The values for these parameters are set by the scientist, and the simulation is then run on a computer.

[24] J. Lenhard, G. Küppers, T. Shinn (eds.) (2006) *Simulation: Pragmatic constructions of reality.* Dortrecht: Springer; T. Shinn (2006) When is simulation a research technology? Practice, markets and lingua franca. In: J. Kueppers, J. Lehnard, T. Shinn (eds.) Simulation: Pragmatic construction of reality. Dortrecht: Springer, pp.187–205.

[25] See in particular the role played by simulation in the cognitive advance and community development of quantum chemistry. K. Gavroglou, A. Simões (2012) *Neither Physics Nor Chemistry*. Cambridge, MA: The MIT Press.

Depending on the perceived credibility of the results, the scientist will subsequently modify particular variables of the model and then rerun the simulation. The aim here is twofold. Based on background knowledge and simulation experience, the practitioner can establish a model of a phenomenon that enhances knowledge of the forces at work and the physical structure. Alternatively, the scientist may strive to produce a model containing specific variables and values which leads to a finding that matches the results of colleagues engaged in physical metrological laboratory experiments on concrete matter—as opposed to virtual matter expressed by idealized findings. The kind of simulation current in nanoscale research did not flower before the late 1980s and early 1990s. This is explicable in terms of the evolution of computer hardware, the emergence of computer graphic programs, and the full development of a variety of analytic algorithms appropriate to the category of problems important to nano.

As indicated at the opening of this discussion, computer simulation involves three complementary components. Electronic hardware constitutes one stream in the process of the rise of simulation in science. The first electronic computers were introduced in the early 1940s in conjunction with the technological problem of finding a solution to detonation of the atomic bomb. The program employed here was stochastic. It was uneconomical in computational time and approximate in results: this was the Monte Carlo program. The computer used was cumbersome and slow, employing thousands of vacuum tubes and mechanical relays. Construction of the pioneering device was necessarily preceded by crucial theoretical and mathematical progress. On the eve of World War Two the British cryptologist, Alan Turing, introduced the concept of what soon became known as the "theoretical Turing machine." Turing advanced the idea that it is possible to resolve any soluble mathematical calculation using a machine equipped with an infinitely long numerical tape and a mobile reading head. In 1937 the American mathematician, Claude Shannon, envisaged the construction of rapid electronic devices based exclusively on relays and switches. Engineers were quick to assimilate this concept and to execute it, leading directly to advances in electronic computation. In 1948 Shannon published his famous paper, "A mathematical theory of communication," which introduced the idea that intelligibility takes the form of informational communication, and that such information and its transfer, precisely match the potential of numerical calculation of the sort entailed in computers.[26]

The first electronic computer employing transistors was built in 1954; this was the grandfather of the modern numerical computer. Thanks to the invention of the integrated circuit and microprocessor, it rapidly became possible to place more than one thousand transistors on a single chip and thus to reinforce computational capacity. In the early 1970s IBM marketed its model "360," a large and for the day powerful yet rather expensive device, and consequently still relatively restricted in circulation. Shortly thereafter, the company manufactured the MTM-70, and the Commodore—the latter in 1981.

[26] C. Shannon (1949) Communication theory of secrecy systems, *Bell System Technical Journal*, 28(4): 656–715.

The development of the microprocessor and the personal computer revolutionized the computer world by enhancing computational capacity, extending the range of applications—notably in science where it was no longer always necessary to have access to big mainframe devices—and by multiplying the number of people who could acquire computers.[27] With reference to science, during the late 1980s, it had become commonplace for scientists to have a computer on the desk for preparation of texts and increasingly for running simulations.[28] Laboratories specializing in simulation emerged. These sometimes had access to huge powerful computers such as the Crey-1, introduced in 1976, and later to the Crey-2 or Crey-3. However, availability to this kind of device was relatively rare in science and notably so in nanoscale research. Finally, the personal computer democratized and extended computation in an additional way. With the emergence of Intranet and Internet, it became possible, and even current, to connect hundreds and even thousands of personal computers such that, when their individual operators were not using them, the computer would automatically be made available to the processing of a part of a sizable simulation program that lay beyond the scope of any single personal computer. This is known as distributed computation.

But computer hardware constitutes only one stream. The development of computer models capable of meshing fundamental equations to models and to physical matter is equally necessary. The development of appropriate software in the form of programs has been similarly important. How to transform the basic quantum mechanics equations associated with the structures and properties of atoms and molecules into soluble problems which can then provide useful information for different kinds of materials.[29] Much of the work on programming specific to scientific research has been rooted in the discipline of chemistry. This is perhaps in part due to the convergence in chemistry of basic investigations and engineering endeavors and research. Moreover, chemistry is by far the biggest community in science. Computational chemistry now constitutes an independent sub-field, which boasts its own professional association and prestigious review.[30]

[27] C. Lécuyer, (2007) *Making Silicon Valley. Innovation and the growth of high tech, 1930–1970*. Cambridge, MA: MIT Press; C. Lécuyer, D. Brock (2010) *Makers of the Microchip. A documentary history of the Fairchild semiconductor*. Cambridge, MA: MIT Press; T. Shinn (2013) The silicon tide. In: R. Fox (ed.), *The Oxford Handbook of the History of Physics*. Oxford: Oxford University Press.

[28] J. Lenhard (2010) Computation and simulation. In: J.T. Klein, C. Mitcham, R. Frodeman (eds.) *The Oxford Handbook of Interdisciplinarity*. Oxford: Oxford University Press, p. 246.

[29] Some nanoscientists whose research is simulation based emphasize the importance of the recently acquired possibility to explore computationally ever greater numbers of molecules or atoms than had been manageable in the past. At first sight, this goal might seem to contradict the often-voiced ambition among most nanoresearchers to focus on the local level, which entails access to only a small number of entities. The project, which involves simulation of many molecules still in a nano perspective, may reasonably be interpreted as an attempt to explore nanoscale objects and relations in the framework of a more extended and perhaps generalized system. C. Noguera, F. Finocchi, J. Goniakowski (2004) First principles studies of complex oxide surfaces and interfaces, *Journal Of Physics: Condensed Matter*, 16(26): 2509–2537. Interview of Claudine Nogura (INSP, Paris) by authors, 23 July 2007.

[30] K. Gavroglou, A. Simões (2012) Neither Physics Nor Chemistry.

The program Gaussian notably incorporates what is known as density functional theory (DFT) and it can also analyze the structure and properties of single molecules.[31] This combination of DFT and single-molecule analysis makes the specialized quantum chemistry simulation program Gaussian a strong contender among computational experimental scientists engaged in nanoscale research. Indeed, DFT is present in almost all of the many programs utilized in nano. DFT permits localization of the electrons present for a given element as specified by the ab initio approach, which operates on the basis of an element's atomic number. It identifies the forces, distances, and geometric relations and interactions between electron clusters in terms of density. Through this analysis of electronic dispositions, simulation provides information about the physical properties of the sample being simulated. Simulators can make predictions about the presence or dominance of properties such as electrical conductivity, mechanical characteristics like tensile strength, friction, etc., and because the simulation program contains specifics concerning which parameters are active and about their numeric values, it can also suggest connections between particular forces, dynamics, and structures, and the existence of particular physical characteristics, in effect, causal relations between structure, dynamics, and material behavior. Because DFT-based simulation can effectively probe the molecular level where nanoresearch often focuses and excels, this potentiality again suits it for nanoscale investigations. Besides Gaussian, other simulation computational programs suited to nano exploration include Against, Absims, Aces, Adf, Adfl, Castep, etc.

Computer graphics constitutes the third and final element central to the emergence of simulation in science and perhaps more particularly to its importance for nano. It is credible to advance this last claim in view of the historical fact that computer graphics has proven crucial to the emergence of static and animated images. And within nanoscale research, it emerges that images occupy a highly central position for the description of materials, for the understanding of relations between structure and behavior of materials, for the communication and grasp of data both inside a specialty and between practitioners of different specialties, and finally for communication between nanoresearch practitioners and extra-scientific communities.[32] Computer graphics, sometimes termed computer-generated imagery, has powerfully affected the three domains indicated above: the entertainment industry (computer games and animated films), engineering in the form of technical design and testing, and finally science. It was the world of film and then technical design that first benefited from simulation graphics, followed by science only in the 1980s and very massively in the 1990s, when it began to penetrate university teaching and became an often routine element of laboratory activities. The concept of employing a computer to draw forms as opposed to perform calculations or write words

[31] J. Lenhard (forthcoming) Disciplines, models and computers. The path to computational chemistry. *Studies in History and Philosophy of Modern Physics*.

[32] M. Ruivenkamp, A. Rip (2010) Visualizing the invisible nanoscale. Study of visualization practices in nanotechnology community of practice, *Science Studies*, 23(1): 3–36; M. Ruivenkamp (2011) *Circulating Images of Nanotechnology*. Twente: Universiteit Twente.

first surfaced at MIT in the 1950s. In the 1970s the computer department at Utah State University became a hive of computer-graphic activity.[33] It is safe to say that it was there that computer graphics came into its own.

While computer-generated imagery penetrated engineering relatively rapidly, scientists turned to images rather late. For many years, much simulation in science depended on numerical tables or matrixes as the vehicle for displaying data. This method of extracting information and developing intelligibility proved both slow and uncomfortable for many researchers. With the improvements in computer-graphic programs and the explosion of computer-visual representations in an increasing range of work, scientists too began to integrate visuals as a platform for studying their objects. Images offered several advantages over a narrowly numerical form of display. As we shall see in Chapter 4, a general perception of results could be obtained at a glance. Moreover, relations between objects were readily perceptible. Computer-based graphics permit scientists to observe structure through the study of forms that witnesses and gives substance to matter via images.

The importance of simulation instrumentation in experimentation in NSR is suggested by the number of laureates of the Feynman Nanotechnology Prize whose work is largely or entirely based on simulation. This highly prestigious prize was founded in 1993 to mark excellence in nanoresearch. It was established by the Foresight Institute, which is one organizational expression of the molecular manufacturing program introduced by Eric Drexler in the 1980s. The Feynman nano prize recognizes achievements in physics, chemistry, engineering, and the life sciences. Between 1993 and 2010 a total of 54 prize winners were designated. Of this number, the research of 29 depended heavily on the utilization of numerical simulation. Charles Musgrave was the first to receive the prize, in 1993, for his work on simulation modeling of a hydrogen abstraction tool useful in nanotechnology. Musgrave was at the time a doctoral student, and the selection of a candidate specialized in simulation may be seen as presaging the future. Four years later, early simulation work on the structure of nanotubes was carried out by Deepak Srivastava and his team, who received the prize in 1997 (see Section 1.2). One of the United States' most experienced and famous simulation experts, William Goddard, was recognized in 1999 for modeling the operation of molecular machine designs. The 2000 prize went to Uzi Landman for his pioneering work in computational materials science for nanostructures, in particular gold nanowires. Work on biological materials constitutes a large group of simulation-oriented Feynman laureates. In 1995, Nadrian Seeman was rewarded for his work, in which he used DNA as building blocks in the construction of nano-architectural structures. In 2004, the prize went to David Becker and Brian Kuhlman for their development of a simulation ground Rosetta program for the design of stable protein structures. In 2005, Christian Schafmeister received the prize for simulation-based experimentation on a novel technology for synthesizing proteins.

[33] E. Francoeur, J. Segal (2004) From model kits to interactive computer graphics. In: S. de Chadarevian, N. Hopwood (eds.) (2004) *The Third Dimension of Science.* Stanford, CA: Stanford University Press: 402–429.

1.2 THE RISE OF NANOMATERIALS

The history of NSR has often been narrated in terms of novel instrumentation. We suggest that the instrumentation narrative demands incrementation. The rise and evolution of nanoscale research is equally a story of the discovery, and above all the synthesis, of new families of low-dimensional physical materials.[34] Whether materials are fabricated by the hand of man or already present in nature, nanoscale practitioners set out to build a research world of objects that are describable, controllable, and which are specifically tailored in order to address pre-formulated questions.[35] This relationship between question and object in which the former determines the parameters of the latter inverts a long standing scientific habitual link where the extant objects of nature dictate the parameters of intellectual inquiry (see Chapter 4).

Crucially important is the fact that nanostructured materials are fundamentally different from the alloys that have figured so centrally in the history of humankind materials. Pre-nano substances, like alloys, are artificial. Alloys such as bronze (copper and tin) and steel (iron and carbon) respectively date back to Antiquity and the seventeenth century.[36] Precipitation alloys like titanium and aluminum are a twentieth-century product, as are reinforcement-metrics composite materials. All these substances share two characteristics: they are bulk materials and they are the result of mixtures. By contrast, nanomaterials are historically novel in three ways: they are characteristically low dimensional (zero, one, and two); their composition is controlled at the atomic level; contrary to alloys where components are interspersed, numerous nanomaterials are fabricated and observable layer by layer.

The emergence of families of novel nanomaterials during the 1980s and 1990s, some of which had their technological origins in the 1970s and even before,[37] seriously transformed the geography of a part of contemporary science. Materials such as buckyballs, carbon nanotubes, nanowires, nanowells and quantum dots, nanocavities, DNA origami, nano Bollean rings, nanomotors, etc. all emerged between 1985 and about 2000. In the pages that follow, we will focus mainly on that part of the nanomaterials revolution

[34] Arthur von Hippel was among the first to propose the idea of engineering at a molecular level. A. von Hippel (1956) Molecular engineering, *Science*, 123: 315–317; A.R. Von Hippel, R.B. Adler, S.C. Brown, C.D. Coryell (1959). *Molecular science and molecular engineering*. Published jointly by the Technology Press of MIT and J. Wiley.; H. Choi, C. Mody (2013) From Materials science to nanotechnology: Institutions, communities, and disciplines at Cornell University, 1960–2000, *Historical Studies in the Natural Sciences*, 43(2): 121–161.

[35] The research of Bernadette Bensaude Vincent on the history of materials science and the epistemology and sociology of technoscience raises important questions about the complex relations between the technology of artificial substances and discontinuity in the evolution of scientific knowledge. See e.g. B. Bensaude-Vincent (2001) The construction of a discipline: Materials science in the United States, *Historical Studies in the Physical and Biological Sciences*, 31(2): 223–248; id. (2009) *Les Vertiges de la technoscience: Façonner le monde atome par atome*. Paris: La Découverte.

[36] <http://en.wikipedia.org/wiki/Alloy>; <http://en.wikipedia.org/wiki/Precipitation_hardening>; <http://en.wikipedia.org/wiki/Composite_material> (20 September 2013).

[37] P. McCray (2007) MBE deserves a place in the history books, *Nature Nanotechnology*, 2: 259–261.

that is referred to as low-dimensional systems and confinement. The notion of "confinement" is often evoked by nanoscientists to define and explain physical properties of low-dimensional systems. Entities such as electrons and photons move along a specific characteristic length and, deprived of this freedom of movement by a restrictive environment, they behave in an unusual fashion that is often rich in physical information. The behavior of a wave/particle depends on the number of degrees of freedom available to it, zero, one, or two, or three as in macroscopic objects. For example, a quantum dot confines a photon of a particular wavelength to change its frequency and energy level in order to generate re-emission. Low-dimensional objects, such as carbon nanotubes, nanowells and nanowires, and quantum dots, are zero-, one-, or two-dimensional substances. A two-dimensional object is a substance that limits the motion of internal components such as photons or electrons to just two dimensions. A nanotube is a rolled sheet of carbon-60. The material is so thin (only one or two atoms thick) that electrons can circulate longitudinally and latitudinally. A one-dimensional object is illustrated by a nanowire, where electrons are exclusively free to move along the length of the wire. Lateral motion is confined by wire diameter. Zero-dimensional objects, a quantum dot, confine electrons in all three spatial dimensions. They do not by and large exist in nature but have been manufactured, synthesized by scientists and engineered, in order to study a range of otherwise inaccessible physical relations and to address novel kinds of questions. Synthesis of novel nanostructured materials has been dependent on the introduction of several recent technologies, including nanolithography, molecular beam epitaxy (see Section 1.3 below), laser ablation, supersonic beams, electric discharge, etc. The elaboration of a new family of carbon-based materials initiated the remarkable expansion of low-dimensional objects and their catapulting of a set of major research fields in nanochemistry and nanophysics. Indeed, for an extended period these materials were often emblematic of nanoscience and nanotechnology.

1.2.1 Pioneering nanomaterials: fullerenes and carbon nanotubes

A new category of carbon was introduced in 1985, which became known as "fullerenes." They were retrospectively associated with the origins of nanoscale research. In nature, there exist three categories of carbon: diamonds, graphite, and charcoal. On heating graphite to over 3000 °C, and then cooling the vapor, scientists observed the condensation of a carbon material whose structure differs radically from that of other forms of carbon. The shape of this substance consists of arrangements of molecules in the form of pentagons and hexagons that constitute a closed carbon cage at the molecular scale. These new materials—fullerenes and buckyballs—were named because of their morphology, which resembles the architecture of a US sports stadium designed by the famous engineer and architect, Richard Buckminster Fuller (1895–1983). A buckyball fullerene can be seen in Figure 1.8. It is a spherical molecule with the formula C60. It has a cage-like fused-ring structure (truncated icosahedron) which resembles a soccer ball made of twenty hexagons and twelve pentagons, with a carbon atom at each vertex of each

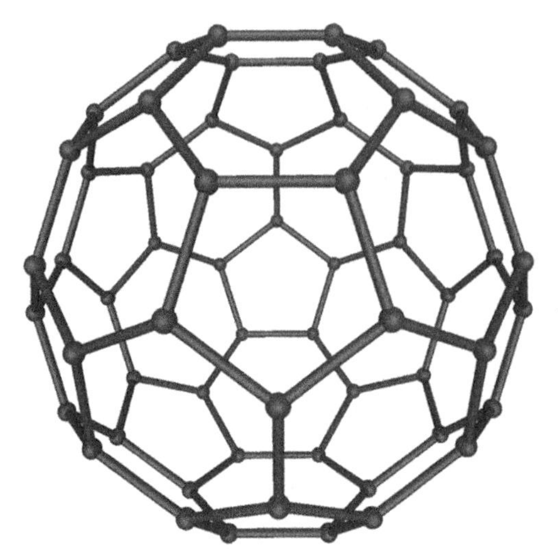

Figure 1.8 Buckyball fullerene. Image created by Michael Ströck, courtesy of Wikimedia Commons.

polygon and a bond along each polygon edge. The identification of fullerene buckyballs immediately stimulated lively interest.

Richard Smalley (1943–2005),[38] Robert Curl (1933–), and Harold Kroto (1939–) synthesized the first fullerene buckyballs in 1985 and received the Nobel Prize in chemistry for their accomplishment in 1996. Smalley was Professor of Chemistry at Rice University. His research in physical chemistry investigated formation of inorganic and semiconductor clusters using pulsed molecular beams and time-of-flight mass spectrometry. Smalley is known as an accomplished instrument designer and maker: fullerenes were discovered in conjunction with the design and construction of a new device for synthesizing materials. The technology consisted of laser vaporization, supersonic fluid dynamics, and a precision spectroscope. Smalley referred to this ensemble as a new kind of microscope. One will see throughout this book that nanoscale researchers are often accomplished and highly original instrument men.

In connection with this instrumentation expertise, Robert Curl, a chemist, introduced Smalley to Harold Kroto, also a chemist who had launched a research program to look for carbon chains in the interstellar medium. The result of this collaboration was, as indicated above, the discovery of C60 (the fullerenes). The research that earned Kroto, Smalley, and Curl the Nobel Prize was communicated in an article published in *Nature* in 1985.[39]

Why was this discovery seen as so important and how does it fit into the origins and evolution of nanoscale scientific research? Firstly, fullerenes constitute a previously unobserved category of carbon. Secondly, their originality is mainly expressed in terms of geometry, and the centrality of geometry in fullerenes constitutes a first step toward the generalization of geometry and forms as a foundation for grasping the nano world.

[38] Richard E. Smalley (1996) Discovering the fullerenes, Nobel Lecture, 7 December.

[39] H.W. Kroto, R.E. Smalley, J.R. Heath, R.F. Curl (1985) C60: Buckminsterfullerene, *Nature*, 318 (6042): 162–163.

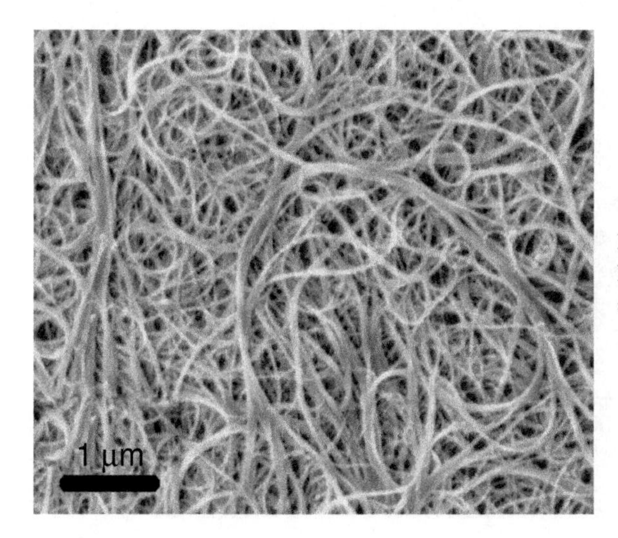

Figure 1.9 Carbon nanotube. Image courtesy of Materialscientist at en.wikipedia.

Thirdly, some exciting new physical properties, such as relations between weight and strength, are exhibited by fullerenes and open the way to interest for synthesizing new categories of materials and studying their physical characteristics. Finally, throughout the 1990s, Smalley emerged as a strong proponent of research at the nanoscale and of the extension and dissemination of nanotechnologies. He argued that fullerenes could be employed in the construction of space elevators that could lift loads 40,000 kilometers into a geostationary orbit or could constitute lightweight storage tanks for fuels. With such ideas, he propagandized the place of nano in the emergence of a kind of new age.

But it was neither the 1985 introduction of C60 nor the discovery of buckyballs that shot nanomaterials to the fore. It was instead the carbon nanotube, synthesized in 1992, that heralded the age of nanomaterials! This substance has been cited in over 89,000 articles between 1992 and 2012. The most-read article has been cited over 18,000 times.[40]

Figure 1.9 evokes the architecture of carbon nanotubes which are cylindrical in shape. They have sometimes been likened to a plate of spaghetti. Nanotubes have been constructed with a length-to-diameter ratio of up to 132,000,000 to 1.

Who introduced carbon nanotubes, and what were the circumstances of their synthesis? Using the electric-arc discharge technique to synthesize fullerenes, in 1992 the Japanese physicist, Sumio Iijima (1939–), produced a C60 architecture different from that of the cages of buckyballs. Iijima called his material "carbon nanotubes." Iijima received his PhD in solid-state physics in 1968 at Tohoku University in Sendai (Japan). Between 1970 and 1982, he conducted research with crystalline materials and high-resolution electron microscopy at Arizona State University. He visited the University of Cambridge during 1979 to perform studies on carbon materials. Iijima's subsequent career principally took place in research institutes in Japan. In 2008, he was the first beneficiary of the Kavli Prize

[40] S. Iijima (1991) Helical microtubules of graphitic carbon, *Nature*, 354 (6348): 56–58.

Figure 1.10 Armchair carbon nanotube. Image from Wikimedia Commons.

for his work on carbon nanotubes—the same year Louis Brus was awarded the prize for his research on the quantum dot (see below).[41]

What are the multiple reasons that resulted in the centrality of carbon nanotubes in NSR? Five factors underpin this situation:

(1) Carbon nanotubes encompass multiple morphological and structural constructs. This entails control. As indicated above, the generic form of the material is cylindrical, having an overall aspect of a pack of spaghetti. This generic form can be conjugated into numerous very different morphologies. Research on carbon nanotubes immediately rocketed. This was due in part to the possibility of synthesizing different architectures—single wall, two walls, multiwalls, etc. Moreover, the configuration of carbon nanotubes takes two fundamental forms: zig-zag and armchair. Figure 1.10 illustrates an armchair carbon nanotube in which one can readily observe the hexagonal structure of the material units and whose shape is reminiscent of an armrest of an easy chair. Alternative architectures and configurations were seen to generate different physical properties, and hence prompted intense investigation. Moreover, synthesis remained problematic.

(2) The substances constitute a remarkably rich terrain for the generation and exploration of new properties. The combination of the rolling angle and radius decides the nanotube's properties; for example, whether the individual nanotube shell is

[41] The Kavli Prize was established in 2005 through a joint venture between the Norwegian Academy of Science and Letters, the Norwegian Ministry of Education and Research, and the Kavli Foundation. The main objective of the prize is to honor, support and recognize scientists for outstanding scientific work in the fields of astrophysics, nanoscience and neuroscience, and to award three international prizes every second year. The Kavli Prize was awarded the first time in Oslo on 9 September 2008.

metal or semiconductor. Mechanical properties include strength, stiffness, and hardness. Because of the symmetry and unique electronic structure of graphene (a two-dimensional crystalline category of carbon), the structure of a nanotube strongly affects its electrical properties. All armchair nanotubes are metallic, and all zig-zag nanotubes are semiconducting. Of high interest, unanticipated and extremely interesting positron behavior is observed in certain geometries of carbon nanotubes.

(3) Carbon nanotubes are considered as a low-dimensional system. The walls of tubes measure only a few atoms or a few nanometers in thickness. This makes them two-dimensional objects. In the shape of flattened sheets, the motion of electrons is confined to paths along the length or the width of the object. The third dimension is confined to its nanometric size. As said above, low-dimensional systems are important in NSR, and NSR is indeed the kingdom of low dimensionality.

(4) They were taken seriously as a terrain for multiple applications, not least of all by the US military and the federal government. Because of their startling mechanical and electronic properties, carbon nanotubes were immediately perceived as rich in potential applications—this beyond their interest in terms of fundamental chemistry and physics. Rather quickly, the amount of research on carbon nanotubes became so massive that the nanomaterial emerged as a kind of subcommunity of research. Industry and national science agencies financed investigation, and nanotubes became the topic of countless science conferences.

(5) Carbon nanotubes also proved central to the development and recognition of the field of computational nanoscience. The episode in which the US Air Force investigated the possible applicability of carbon nanomaterials to materials intended for space exploration is illustrative of the development of the enduring link between simulation-based experiments and nanomaterials.

In view of the perceived importance of carbon nanotubes for learning and for applications, The National Aviation and Space Agency, NASA, began a huge research project in 1997, and correspondingly developed a special nanocomputation science section, under the direction of the already famous Deepak Srivastava (Director of the NASA Computational Nanotechnology Laboratory). As indicated above, in 1997, Srivastava received the Feynman Nanotechnology Prize for his theoretical contributions. His research group, which focused its efforts on the study of carbon nanotubes, was based at the Ames research facility in California.

Srivastava's work on carbon nanotubes was intended to determine the macroscopic characteristics of the material in the light of molecular and atomic structures and dynamics.[42] Put differently, he investigated the impact of the physical and electronic interactions of molecules on the strength of bulk objects. It proved extremely difficult to determine

[42] A. Johnson (2006) The shape of molecules to come. In: G. Küppers, J. Lenhard, T. Shinn (eds.) *Simulation: Pragmatic Construction of Reality*. Dordrecht: Springer, pp. 25–39.

the relationship between microscopic characteristics and the way bulk materials functioned. Each scale—atomic, molecular, and macroscopic—could effectively be explored separately. It remained to be seen, however, how atomic structures affected molecular behavior and how molecular behavior characteristics influenced the behavior of the object itself. At that time this thorny problem was best explored through computational operations: these were early days for multi-scale modeling.[43]

This powerful possibility is amply documented by the results of Srivastava's simulation work on carbon nanotubes. It is to be noted that such complex relations were not accessible to metrology-based experimentation. Implementation requires momentous computational capacity to obtain precise results. The NASA work on the properties and the potential applications of carbon nanotubes was given high priority, to the extent that Srivastava enjoyed access to the United States' most powerful computational hardware. Between 1998 and 2002, he worked with the SGI Origin 2000 supercomputer and could call upon distributed-calculation systems. These were the best the nation had to offer. The research of the US scientist, Deepak Srivastava, illustrates the conditions under which carbon nanotubes so quickly became cognitively, materially, and institutionally important, and it indicates their scientific and engineering significance.

The trajectory of carbon nanotubes corresponds to what is perhaps a more general geography. In the case of the investigation of carbon nanotubes, it involved the development of new technologies. Note that the essential point of this new material lies in its internal geometry and form, which are responsible for the generation of particular physical properties. Extensive research followed on the synthesis of this material and involved large numbers of scientists. This was accompanied by abundant metrological experimentation on the physical properties of carbon nanotubes, which was often tied to and anticipated by simulation.

1.2.2 The quantum dot

What is a quantum dot and why is discussion about it important to a historical and epistemological appreciation of NSR? A quantum dot is one category of crystal. Crystals fall into three categories: coarse (10,000 to 2500 nm), fine (2500 to 1000 nm), and nano (100 to 1 nm). A quantum dot is a subcategory of nanocrystals that measures less than 30 nm.[44] Scientists are interested in quantum dots because it is known that they possess special optical and electronic properties. Scientists have been similarly interested in them because they do not frequently (if at all) occur in nature, and therefore practitioners have pursued investigation of their synthesis. In turn, effective synthesis itself requires additional knowledge grounded on fundamental research. Investigation of properties thus

[43] E.Winsberg (2006) Handshaking your way to the top: simulation at the nanoscale. In: G. Küppers, J. Lenhard, T. Shinn (eds.) *Simulation: Pragmatic Construction of Reality*. Dordrecht: Springer, pp. 139–155.

[44] C.G. Granqvist, R.A. Buhrman (1976) Ultrafine metal particles, *Journal of Applied Physics*, 47: 220–2219; L.B. Kish, J. Söderlund, G.A. Niklasson, C.G. Granqvist (1999) New approach to the origin of lognormal size distributions of nanoparticles, *Nanotechnology*, 10: 25–28.

must be preceded by research on the production of the material itself. NSR entails dialog, and beyond dialog even intertwining between two levels of exploration: synthesis and phenomenology.

Many fundamental properties of quantum dots depend upon the solid being periodic over a particular length scale. Stated differently, the crystal must be internally homogenously organized because, if there is a deviation, its characteristics change, as does its performance. In view of the size and architecture of these materials, and the internal forces associated with these objects, physicists and chemists can study phenomena specifically linked to the crystal's size, form, ionization potential, melting point, band gap, and magnetic saturation. Here, quantum dots serve as vehicles for the exploration of defined and important physical dynamics. They constitute a controlled physical environment favorable to the investigation of precise questions. Quantum dots are specially designed and tailored for a given experiment. Their synthesis constitutes one stage of the research process and indeed a critical one. As will be seen in Chapter 2, much research cannot be carried out in the absence of quantum dots, and other like materials, specifically designed and built for exploration of the selected question.

There is another important connection between materials research and physical research that stands out in NSR: some synthesized substances exhibit properties that make them particularly amenable to research transversality. They can generate original properties across a range of different scientific questions. This section will examine the ways in which quantum dots emerged and developed. The discussion will suggest why quantum dots are often so central to diverse domains of NSR.

The history of the quantum dot is complex, involving multiple actors. It was discovered during the early 1980s.[45] Louis Brus was one of the early fathers of the quantum dot. He received the prestigious nanoscience and nanotechnology Kavli International Prize in 2008 for his early efforts. His investigations were an extension of his previous research on quantum wells, which are two-dimensional systems. A quantum well can be pictured as an enclosure into which electrons flow, down like water into a well. A quantum dot is an entirely confined zero-dimensional system where freedom of movement of electrons is absent. In quantum dots, electron confinement is paralleled by confinement of photons; photons take on unusual properties resulting in intense emission of light and change of color. Emission wavelength-enhanced color of light is size dependent. This has technological implications.

In the 1970s, while working at the AT&T Bell Laboratories, Brus and others contributed to the birth of the quantum dot in a framework of evolution in micro-electronic components and basic research on crystallites and electron-confinement effects.[46] Brus

[45] In the early 1980s, the Soviet solid state physicist Alexey J. Ekimov observed a strange behavior of a crystal in a glass matrix. He hypothesized the possibility of a nanometric crystal whose existence in nature remains problematic and unresolved. Ekimov is frequently associated with the discovery/synthesis/optical properties of nanocrystals (see <http://en.wikipedia.org/wiki/Alexei_Ekimov>, and <http://en.wikipedia.org/wiki/Quantum_dot> (consulted 28/09/2013).

[46] L. Brus (1984) Electron electron-hole interactions in small semiconductor crystallites. The size dependence of the lowest excited electronic state, *Journal of Chemical Physics*, 80(9): 4403–4409.

was one of several scientists engaged in research on low-dimensional systems and confinement during these years. An abundance of studies were being undertaken along these lines. Mark Reed, at Texas Instruments, introduced the technology for synthesis of quantum dots in 1987. Many other physicists, among them Michael Steigerwald at Bell Labs and Paul Alivisatos at the University of California, Berkeley Campus, were involved in these investigations.[47] People were inventing novel materials in order to answer novel questions. Highly original fundamental research had to be undertaken to produce the wanted materials.

The work of the famous physical chemist, Paul Alivisatos (1959–), points to the vast range of questions and activities in the area of colloidal inorganic quantum dots. Alivisatos earned a PhD in Physical Chemistry under the supervision of Charles Harris from the University of California, Berkeley, in 1986. He was a postdoctoral fellow at Bell Labs with Louis Brus. He is presently Professor of Chemistry and Materials Science and Professor of Nanotechnology at Berkeley. His research includes problems such as scaling laws linked to temporality, isomerization of a crystal between two stable bonding geometries, the relationship between crystal size and the absorption and emission of light, what constitutes the largest crystal that can be made defect-free by precisely controlling the size and surface of a nanocrystal, and how its properties can be tuned and controlled. Alivisatos' work is interesting for documentation of the relationship between research questions and the specificities designed into research materials.

The ability to synthesize quantum dots of high quality is omnipresent in the vast majority of Alivisatos' research. As indicated above, this is due to the importance of the precise control necessary for obtaining the desired size, shape, and surface structures of nanocrystals. Indeed, in nanoscale research in the physical sciences, synthesis constitutes a sometimes separate and always important field of study in its own right. Alivisatos is a master of the synthesis of colloidal inorganic nanocrystals through injecting surfactants. He also employs molecular beam epitaxy and DNA hybridization. With quantum dots, Alivisatos explores three principal domains:

(1) Optical phenomena related to crystal light emission. Here, optical phenomena are explored using quantum dots ranging between 2.6 and 6 nm. Alivisatos has detected confinement properties that favor emission of light in the blue spectrum. One significant conclusion concerns the difference in behavior of such quantum dots when synthesized by different technologies: molecular beam epitaxy (see below) and surfactant technology. The intertwining of different techniques of synthesis and physical properties reinforces the centrality of synthesis, linking it almost organically to experimental findings.[48] This suggests that, for a given material, there also exists a "materiality" that produces specific effects and whose

[47] M. Reed (1993) Quantum dots. Nanotechnologists can now confine electrons to pointlike structures. Such "designer atoms" may lead to new electronic and optical devices, *Scientific American*, 268(1): 118–123.

[48] A.A. Guzelian, U. Banin, A.V. Kadavanich, X. Peng, A.P. Alivisatos (1996) Colloidal chemical synthesis and characterization of InAs nanocrystal quantum dots, *Applied Physics Letters*, 69.

particularities are often ill described and incompletely understood (see Chapter 4). The concept of materiality of materials may in itself emerge as a new inquiry within nanomaterial science.

(2) Electron phenomena linked semiconductors and band-gap as expressed in transistors. Preliminary studies by Alivisatos and colleagues demonstrate that both the electrical and the optical properties of individual quantum dots clearly show that a single excess charge on a quantum dot can markedly influence its properties. Measurements are performed on electrical transport in a single-electron transistor made from a colloidal nanocrystal of cadmium selenide. This device's structure enables the number of charge carriers on the quantum dot to be directly tuned, and so permits determination of the energy required for adding successive charge carriers. Such measurements are invaluable for understanding the energy-level spectra of small electronic systems.[49]

(3) Inorganic quantum dots as biomarkers. Alivisatos' research on inorganic quantum dots has carried him into the world of molecular biology, where crystals can function as markers. In the area of quantum dots as biological conjugate tag molecules, research focuses on the adaptation of quantum dots to the biological environment, which itself constitutes a major research problem. Here quantum dots are used as sensing devices when connected to biological materials such as DNA or proteins (see Chapters 3 and 5).[50] Through their light-emitting capacity, they allow biologists to observe the length, folding, and displacement dynamics of proteins, and can be used for example to detect cancers. Here Alivisatos' work consists of synthesizing quantum dots specifically adapted to biomaterials. This is a strong example of the relationship between a research question and a material by design

Finally, as implied above, for Alivisatos, synthesis of nanocrystals constitutes a necessary terrain for addressing some important theoretical questions in solid state chemistry. For example, what governs the rate of growth of a crystal and what affects its internal and external structures? Many of Alivisatos' publications explicitly discuss the kind of synthesis he employed on the reported project, and details are often presented.

1.3 TECHNOLOGIES OF NANOMATERIALS

Research and fabrication practices presented under the rubric of "synthesis" constitute a sizable portion of NSR-related publications. Some articles deal with techniques for making materials that are the basis of fundamental research, where others are associated with routinized and standardized production of commercial substances. We will present two

[49] D. Klein, R. Roth, A. Lim, P. Alivisatos, L. McEuen (1997) A single-electron transistor made from a cadmium selenide nanocrystal, *Nature*, 389: 699.

[50] P. Alivisatos (2004) The use of nanocrystals in biological detection, *Nature Biotechnology*, 22(1): 47–52.

orientations for producing such nanostructured materials. Both of them emphasize the role of control in their synthesis. The first, molecular beam epitaxy (MBE), combines highly complex instruments and the organization of almost exclusively semiconductor samples. The second, self-assembled monolayers (SAMs), uses natural properties of certain chemical substances as the motors of self-assembling and organization.

1.3.1 Molecular beam epitaxy

Epitaxy is critical to the synthesis of many nanostructured materials. The world of epitaxy synthesis, the structure of epitaxy objects, and the perception and research projects on these materials are central to the post-1990 acceleration of NSR. The term "epitaxy," from the Greek *epi*, "on" and *taxis*, which means "in an ordered rank," may be translated as "arranged on top of." In the context of NSR, it refers to the mastering and fabrication of "artificial" materials that thus do not occur in nature, and are characterized by their size, shape, and by precisely predetermined properties ("materials by design"). In order to create these materials, epitaxy techniques accelerated, developing rapidly in the 1960s. MBE techniques fabricate highly controlled atomic- or nanometric-thick layers of materials that are deposited molecule by molecule one on top of the other. The ultimate structure of the epitaxial substance is determined by the super-lattice substrate, where each film is organized periodically and where the organization between layers is also periodical. This means rigorous control at the atomic level. The resulting nanometric object often exhibits properties absent from natural nanometric materials and bulk substances.

For its part, MBE is connected with both the tradition of thin layers and crystal growth. It is this combinatorial that constitutes the logic and methodology of MBE, that entails particular combined sets of technology, and that offers materials with structures well suited to NSR. Studies of thin films enjoy a long tradition dating back to the early twentieth century.[51] In the 1930s, Erving Langmuir (1881–1957), Nobel Prize in chemistry in 1932, specialized in surface chemistry obtained atom-thick surface layers. During the same period, German specialists in the domain of crystal growth, Max Volmer (1885–1965), Ivan Nikolov Stranski (1897–1979), and colleagues, developed different procedures for controlling the dynamics and structures in the formation of crystal layers.[52] The Stranski and Krastanov technique is still sometimes used for the synthesis of quantum dots.[53] A conjunction between thin film and crystal growth contributed, during the 1950s and 1960s, to micro-electronics, as expressed in the production of transistors and

[51] G. Wulff (1901) On the question of speed of growth and dissolution of crystal surfaces, *Zeitschrift für Krystallographie und Mineralogie*, 34(5/6): 449–530; L. Vegard (1921) The constitution of the mixed crystals and the filling of space of the atoms, *Zeitschrift für Physik*, 5(April–July): 17–26 (doi: 10.1007/BF01349680); A. Reuss (1929) Account of the liquid limit of mixed crystals on the basis of the plasticity condition for single crystal, *Zeitschrift für Angewandte Mathematik und Mechanik*, 9: 49–58.

[52] L. Hoddeson, E. Braun, J. Teichmann, S. Weart (eds.) (1992) *Out of the Crystal Maze: Chapters from the history of solid-state physics*. Oxford: Oxford University Press.

[53] N. Stranski (1928) On the theory of crystal accretion, *Zeitschrift für Physikalische Chemie–Stochiometrie und Verwandtschaftslehre*, 136(3/4): 259–278.

microprocessors. Demands for these components served as a whip for the development of new production technologies.[54] What has become the most prevalent and innovative epitaxy technology, MBE, has been described and analyzed in a seminal article by Patrick McCray.[55]

In 1968, Arthur Y. Cho (1937–), working at Bell Laboratories, invented the molecular beam epitaxy technology for the highly controlled growth of atomic-dimension thin films on super-lattice substrates. Inducement for the invention of efficient epitaxy capacities was largely linked to a commercial demand for more routinized manufacture of the miniaturized semiconductor materials that had become fundamental to much of modern electronics. During the 1970s, Cho's system became central to transistor and microprocessor manufacturing and, in 1980, it was adapted to the production of metallic thin films. On another register, MBE strongly affected experimental quantum physics. Cho's technology has permitted the production of low-dimensional materials such as quantum dots, nanowires, nanowells, and later on, nanotubes.

Cho himself describes the process:

> *In the late 1960s, as devices were getting smaller and smaller, there was a great demand for a crystal growth technology that could prepare single crystalline films as thin as 500–1000 angströms. There was a need to invent a new process. Invention sometimes happens when one combines the knowledge of two established technologies and applies them to a third, to create a new technology. Such is the case for the development of molecular beam epitaxy. Molecular beam epitaxy borrowed the knowledge of surface physics and ion propulsion technology to create a new crystal growth technology. The term molecular beam epitaxy (MBE) was first used in 1970 after several years of extensive studies of atomic and molecular beams interacting with solid surfaces.*[56]

MBE revolves around three technologies: molecular beam physics, ultra-vacuum devices, and ultra-low temperatures (cryogenics). The introduction of MBE may be interpreted as a condensation of these three strands. The first strand, molecular beam physics, arose at the beginning of the twentieth century and is marked by the research of Hans Jürgen Kallmann (1908–91), who, in 1921, was interested in dipole moments and the deflection of beams of polar molecules in an inhomogeneous electric field. This line of research was brought to fruition in 1939, when Isidor Rabi (1898–1988), Nobel Prize 1944, invented a molecular beam magnetic resonance method, in which two magnets placed one after the other create an inhomogeneous magnetic field. This contributed importantly, some twenty years later, to the development of nuclear magnetic resonance.[57] This signifies control over the emission and diffusion of selected materials in the form of molecules and in a gaseous phase.

[54] C. Lécuyer (2007) Making Silicon Valley; C. Lécuyer, D. Brock (2010) Makers of the Microchip; T. Shinn (2013) The silicon tide.

[55] P. McCray (2007) MBE deserves a place in the history books.

[56] A.Y. Cho (1999) How molecular beam epitaxy (MBE) began and its projection into the future, *Journal of Crystal Growth*, 201: 1–7.

[57] C. Reinhardt (2006) *Shifting and Rearranging. Physical methods and the transformation of modern chemistry.* Sagamore Beach, MA: Science History Publications.

A second contributing technology consists of ultra-high-vacuum apparatus. In MBE, these perform two functions: firstly, high vacuum eliminates water vapor and chemical impurities that otherwise contaminate the intended crystal; secondly, it contributes significantly to the environment required for complete control of molecular deposition. Beyond material science, generation of ultra-high vacuums is central to progress in many areas of research. Today it constitutes an autonomous field of instrument research, hosting the production of thousands of articles and possessing highly prestigious journals (for example, the *Journal of Vacuum Science and Technology*, created in 1964).

The final strand, ultra-cold, also dates back to the early twentieth-century research of Heike Kamerlingh Onnes (1853–1926), 1913 Nobel Prize in physics.[58] His devices and technology approach the production of absolute zero. Cryogenic technology quickly became crucial to much experimental quantum-oriented research, such as superconductivity and superfluidity. It has also been linked to many technologies and productions, such as the hydrogen bomb. The key journal, *Cryogenics,* was created in 1960. Ultra-cold is essential to several areas of material science. In the case of MBE, a cryogenic environment is essential, as it offers a thermal sink for the removal of impurities that cannot be eliminated by ultra-vacuum. In this flurry of activity, the domain of epitaxy acquired a supplementary dimension in the form of the perspective of thin film which changes the landscape and environment of crystal growth.

The three above-described strands come together to form the huge and highly complex instrument depicted in Figure 1.11 where a forest of multi-sized cylinders, tubes, and cables combine.

MBE of thin films involves five operations: (1) vaporization of targeted materials such as arsenic and/or gallium; (2) computer-controlled closure of the chamber that houses the substrates on which the vaporized atoms are deposited; (3) careful regulation of atomic or molecular deposition on the substrate, controlled by the creation of an ultra-vacuum; (4) a super-cooled environment set at 77 Kelvin, for the evacuation of impurities; (5) an electronic detection system for purposes of identifying the thickness of layer and control of the entire process.[59]

It would be misleading to imply that MBE dominates the domain of epitaxy. Its position in the epitaxy niche is very specific. Between 1971 and 2013, the ISI Web of Science reports about 44,000 publications connected with MBE. For the period 1964 to 2013, over 80,000 articles were published that highlighted the alternative epitaxy technology of chemical vapor deposition (CVD). The two technologies are often associated with contrasting functions and communities. CVD is generally linked to epitaxial activities touching on engineering and commercial industrial production. Materials are mass-produced and often standardized. The objective is high efficiency. MBE, on the other hand, tends to be located in research facilities that are specialized in the production of research-objects designed and built for the investigation of designated physical properties or dynamics.

[58] K. Gavroglou, Y. Goudaroulis (1989) *Methodological Aspects of the Development of Low Temperature Physics, 1881–1956: Concepts out of context(s)*. Dordrecht: Kluwer Academic Publishers.

[59] Reflection high energy electron diffraction (RHEED).

Figure 1.11 Molecular beam epitaxy. Image reproduced with permission from Professor John Foord, University of Oxford.

MBE equipment is expensive and frequently difficult to operate. Mass-production is not the aim. Custom-made single batches represent the mode of production and the mentality of MBE. Indeed materials are tailored for the requirements of specific experiments (see Chapter 2). MBE personnel are generally scientists or extremely high-level engineers, and many of these scientists are engaged in their own research in parallel with supplying materials for other investigators. In short, MBE is often an integral part of the nanoresearch process, as opposed to the more current form of epitaxy, CVD, which is often industry oriented.

Since its origins, MBE has increasingly allowed synthesis of a range of semiconductor materials, the control of whose size and form offers incomparable possibilities for the exploration of electronic optical and magnetic properties that often surpass the constraints associated with "natural" materials. Such semiconductor materials have also proven significant for research in surface physics and, notably, one can point to investigation of very precisely tailored semiconductor/oxide substances (see Chapter 2).

1.3.2 Self-assembling monolayers

Self-assembling monolayers (SAMs) are an additional entirely novel and highly important material in NSR. Their contribution to nanoscale investigation is crucial because they are

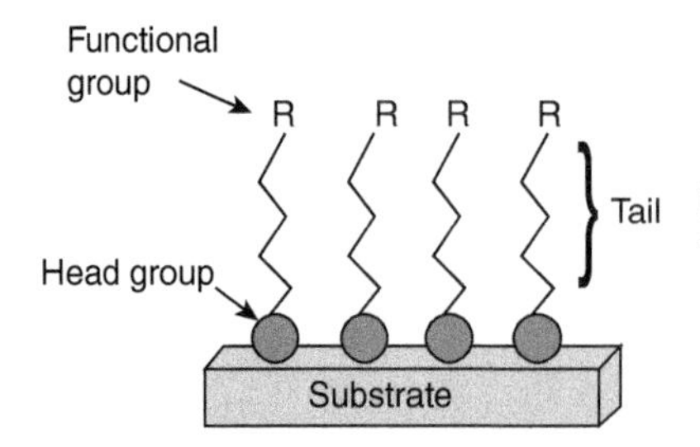

Figure 1.12 Self-assembling monolayer. Image from Wikimedia Commons.

often necessary to the exploration of other substances and because they constitute an object of keen interest in their own right, in view of their little-understood properties.

SAMs appeared during the same decade as buckyballs, carbon nanotubes, nanowells and nanowires, quantum dots, the STM, and the AFM. As will be shown below, the materials of self-assembling monolayers function mainly through chemical dynamics, as opposed to the principally physical dynamics entailed in epitaxy procedures.

A SAM is a self-organized layer of molecules in which they adopt a defined arrangement without guidance or management from an outside source. The only controlling elements are the properties of the environment and substrate. These include chemical properties such as alkalinity, acidity, temperature, electric field, etc. The substrate does not necessarily cover an entire surface. A substrate may be deposited as islands or figures, and it is uniquely there that SAMs grow. The pattern that SAMs effect on the substrate can be designed by tailoring the substrate's geometry. Here and throughout nano, the question of form is of foremost importance.

As shown in Figure 1.12, the molecules that compose the monolayer possess a "head group," which adheres to a specific surface (the substrate), and a "tail." To the end of the tail is attached the functional group, which attracts and fixes the target nanoparticles due to their respective chemical affinities. The attachment of the head group of each molecule of the monolayer is reversible. For example, if one changes the environment of the substrate, the monolayer can detach. In this way the SAM functions as a switch. SAMs incorporate a great variety of properties, which in part are linked to the angle between substrate and molecules, molecular torsion, etc.

The first publication concerning SAMs appeared in 1984, by Jacob Sagiv.[60] Sagiv is today based at the Weizmann Institute in Israel,[61] where he continues to work on different aspects of self-organizing monolayers. He was among the first to develop and utilize SAM-based nanolithography, which has considerably extended the cognitive and technological possibilities of SAM material. The year 2000 saw the publication of over 1000 articles, and by 2013, the total publications has risen to 31,000.

[60] R. Maoz, J. Sagiv (1984) On the formation and structure of self-assembling monolayers, 1: A comparative ATR-Wetability study of Langmuir-Blodgett and absorbed films on flat substrates and glass microbeads, *Journal of Colloid and Interface Science*, 100(2): 465–496.

[61] <http://www.weizmann.ac.il/> (consulted 4/09/2012).

What functions do SAMs perform in nano? (1) They are a vehicle for transporting specified nano objects. (2) They deliver objects to well-defined locations. (3) They can be activated and deactivated countless numbers of times. Taken together, these three characteristics comprise a consummate control mechanism on the nanoscale, and beyond, for a huge quantity of transport compounds and targets.

This control function is not an end in itself, however. It is instead a means intended to promote a cognitive aim. As will be pointed out in the following chapters of this book, molecular form is of the utmost importance in NSR: it both determines numerous properties and is employed as a source of information in observing and communicating. The control potential latent in SAMs is a principal agent for determining the form of material on the nanoscale. Examples of this abound. In one important experiment, the molecular structure of a certain category of SAMs was employed to provide the architectural structure of single-walled carbon nanotubes.[62] Using SAMs, these objects can be duplicated an infinite number of times, which allows standardized observation in a variety of environments. In other instances, SAMs have been mobilized to generate numerous different patterns of the same substance. Using slightly different SAMs, the molecules of the material can be patterned horizontally or vertically. Through such control, experimenters can observe the contrast in properties associated with contrast in forms.[63] In such scenarios, in NSR the importance of materials goes beyond the creation of artificial substances; as in the case of SAMs, it also entails crafting materials in such a way that the historical boundary between an object and experimentation on it is somehow modified. One has moved here from a passive relationship between object and exploration to a historically new, more interactive relationship.

CONCLUSION

The emergence of nanoscale research has modified significant profiles of scientific investigation. It is the emergence of new categories of metrological instrumentation and their linkage with numerical instrumentation, and the capacity to synthesize types of materials that do not exist in nature, that constitute the cornerstone of nanoscale research. From these twin pillars have evolved a series of connected consequences. Four aspects have been notably affected. Firstly, the introduction of scanning probe microscopy in a variety of forms during the mid-1980s made it possible to explore matter in a detailed way on the molecular level. Scanning tunneling microscopy and atomic force microscopy allow the exploration of matter in terms of single atoms and molecules. While earlier

[62] Y.H. Wang et al. (2006) Controlling the shape, orientation, and linkage of carbon nanotube features with nano affinity templates, *Proceedings of The National Academy of Sciences of the United States of America*, 103(7): 2026–2031.

[63] B.J. Vesper, K. Salaita, H. Zong, C. Mirkin, A. Barrett, B. Hoffman (2004) Surface-bound porphyrazines: Controlling reduction potentials of self-assembled monolayers through molecular proximity/orientation to a metal surface, *Journal of the American Chemical Society*, 126(50): 16653–16658.

achievements in electron microscopy permitted the localization of molecules or even atoms, it now became possible to study their shape and internal structure, and to localize entire collectives of molecules. An additional species of instrumentation, namely computational instrumentation, began to flourish during the 1990s, and perhaps more than in any other field of research, interlacing arose between computational instrumentation and metrological instrumentation.

Secondly and equally important, nanoscale instrumentation has allowed scientists to control single molecules and atoms in terms of their position and sometimes interconnections with other molecules. Molecular control has thus become a cornerstone of nanoscale research. This indeed has opened a new chapter in the history of the physical and biological sciences.

The third aspect has been epistemological. In nanoscale research the understanding of matter is based on a deterministic epistemology, as opposed to a strictly statistical interpretation, which was largely the case in recent decades. The importance of determinism is connected to the capacity to observe and study single atoms and molecules, and the ability to control behavior.

The fourth element resides in the fact that nanoscale research is a product of the introduction of entirely new categories of materials. This process began with the synthesis of fullerenes in the 1980s and 1990s. In NSR, it became possible for material scientists to control the deposition of atomic and molecular layers almost atom by atom. Very quickly, the mastery of new methods in nanoscale research materials gave rise to a wave of new substances in the form of carbon nanotubes, nanowells, nanowires, quantum dots, nanoparticles, etc. It was the introduction of new species and combinatorials of instruments and materials in the 1980s and 1990s that marked the genesis of nanoscale scientific research, which demonstrates that the field entails considerably more than issues uniquely of size. Seeing small is thus not equivalent to exploring it.

2
Worlds of Nanophysics

In the physical sciences, research is often characterized by a relationship between metrological experimentation and theory, where the scenarios can be either complementary or tense. In NSR, theory is circumscribed; it is instead a descriptive science where two tandems prevail. One consists of a relation between metrological experimentation and epitaxy; the other resides in the relation between metrological investigations and simulation experimentation. Three characteristics of nanophysics will be described in this chapter. Section 2.1, "The gold rush," documents the speedy and sometimes radical thematic shift often suddenly adopted by practitioners on discovering the possibilities of the nanophysics instrumentation revolution of the early 1980s, and the material synthesis possibilities of the following decades. The trajectories of five practitioners who became nanophysicists between 1983 and 1998 are presented. In parallel, two examples are given of a collective drift toward the extension and institutionalization of nanophysics.

Section 2.2, "Materials, research-objects, and physical phenomena," explores the progressive transformation of the samples demanded by experimenters in the course of their research process and the efforts of epitaxiors to respond. Epitaxiors customize—tailor—substances in order to generate specific wanted properties. This is what we term "research-objects." Research-objects are to be distinguished from the synthesis of materials (see Chapter 1), where the latter constitutes broad generic families of matter, such as fullerenes, nanowires, nanowells, and quantum dots. Though the development of a research-object is related to the synthesis of materials, it nevertheless comprises a distinct activity, consisting of a substance's adaptation in response to experimental exigencies. The dependence relationships between experimenters and the epitaxiors who design and produce research-objects can be complicated. Exploration of two detailed case studies illustrates how the development of a research-object leads to the successive reformulation of an initial research question. This dialog between experimental design and design of research-objects is a structuring characteristic of NSR.

The third and closing section of this chapter, "Metrological experimentation and simulation-based experimentation: between difficult encounters and synergy," examines profiles and processes of experimentation for metrology-based experimenters on the one hand, and for simulation-based experimenters on the other. It also explores collaborations between the two approaches. The temporalities of the two groups are importantly different, affecting the ways that each perceives and formulates research questions and what counts as interesting and what is valid. We then discuss the presence of shared characteristics. Simulators and metrology-based experimentation are both essentially underpinned by description where the parameter of form is of foremost analytic significance. This terrain is highly synergistic and it again substantiates the claim that in many significant ways NSR is a combinatorial science.

2.1 THE GOLD RUSH

In 1849, news suddenly circulated that rich gold deposits had been located in the region surrounding San Francisco, and during the ensuing months and years tens of thousands of prospectors swarmed to the area in search of riches; and numerous were those who satisfied their objectives. A similar such gold rush occurred in the Yukon in the late 1890s, when masses of men immediately modified their professional trajectory and adopted a new path and perspective in response to fresh possibilities in a pioneering landscape.[1]

In the domain of scientific research, the massive and almost immediate rush by scientists, during the 1980s and 1990s, toward novel instrumentation (scanning tunneling microscopy—STM) and new materials (buckyballs, carbon nanotubes, the low-dimensional substances—nanowells, nanowires, and quantum dots) suggests parallels with the gold rushes of history.[2] The invention of the STM in 1981 and its crowning with a Nobel Prize just five years later, in recognition of its potential as a radically new category of metrology, is in itself an amazing gold-rush-like episode. Within the span of a decade, hundreds of researchers had abandoned their previous research instruments, and often their former research questions, in favor of the new family of scanning probing device and instrument-related questions. The same gold-rush dynamics marked the syntheses of carbon nanotubes in 1992. This nanoscale material was at once regarded by countless practitioners in a sweep of numerous and diverse fields as offering the possibility to explore uncharted cognitive territories. They quit their earlier lines of investigation and rapidly adopted elements of the nanoresearch perspective that privilege single objects, research by design, control and instrument, and cognitive combinatorials.

The NSR gold rush is similarly discernible in two institutional and collective grassroots-movement initiatives. The older 1980s molecular manufacturing science, technology, and industry initiative developed by Eric K. Drexler and his Foresight Institute, founded in 1986,[3] quickly recognized the cognitive potential and important early achievements of

[1] J. Rawls, R.J. Orsi (eds.) (1999) *A Golden State: Mining and economic development in Gold Rush California*. Berkeley/Los Angeles, CA: University of California Press (California History Sesquicentennial, 2); P. Berton (2001) *Klondike: The last great gold rush (1896–1899)*. Toronto: Kindle Edition.

[2] The term "gold rush" is used here as a metaphor that refers to the displacement of a mass of people in a particular direction for a particular end. In the case of NSR, the object was certainly not monetary: recall that the US National Nano Initiative, with its offer of generous finance, was announced only in 2000, and the gold rush to which we refer occurred between 1985 and 1995 (cf. Introduction).

[3] P. McCray (2012) *The Visioneers: How a group of elite scientists pursued space colonies, nanotechnologies, and a limitless future*. Princeton, NJ: Princeton University Press; C. Milburn (2010) *Nanovision: Engineering the future*. Durham, NC: Duke University Press Books.

nanoscale research, and in view of establishing links with nano, the Feynman Nanotechnology Prize was set up by molecular manufacturing movement partisans in 1993; this rewarded excellence in experimentation and theory in nano investigation.[4] This prize may be viewed as a kind of permanent anchor in the then unmapped and unsettled land of the nano territories. Similarly, in 1992 the Canadian chemist, Geoffrey Ozin, published a widely read, now classic text inciting scientists to engage in the adventure of developing nanostructure materials, a new domain abounding in creative possibilities. The article written by Ozin, which called for the implanting and growth of a large nanoresearch community, remained the most cited text of the prestigious journal *Advanced Materials* for over a decade.[5] It induced a collective movement in favor of nanoresearch—a grassroots mass rush.

In this section, the gold-rush effervescence that occurred at the very outset of nanoscale research, and that often still persists today, will be documented. Attention is drawn to the speed at which the new nanometric instrumentation was adopted. In the early years, practitioners rushed to determine precisely what the instruments consisted of and how they worked, how they could be improved, and what they could do and not do. Other practitioners immediately employed the devices on just about any substance or force dynamic that came to hand. Following a similar logic, as novel nanostructured materials were synthesized, practitioners streamed to them to explore their composition or to determine their structure or properties. Each new substance generated novel expressions of materials or even new kinds of materials, so a circular synergy soon emerged. Finally, the two above-mentioned gold rush institutional/collectivist landmarks, the Feynman Prize and the Ozin project, will be discussed as stabilizing referents in what was an often turbulent moment in the early history of nanoscale science.[6]

James Gimzewski is a Scottish-born physicist, presently Professor of Physics at the University of California in Los Angeles. He was trained in solid-state science during the 1970s. In the mid-1980s he obtained a research position at the IBM Zurich research center, where he was attached to the extended group that had so recently developed the STM. Gimzewski at once engaged in nanoscale studies, the field that he continues to explore. Gimzewski's swift, spontaneous, and total involvement in NSR elucidates numerous aspects (instrumentation, diversified research substances/properties, single molecules, atomic and molecular control) of the nano gold rush. The various activities of Gimzewski convey the excitement, enthusiasm, and energy that surrounded early nano and that continue to animate many practitioners even three decades later.

[4] The prize awarded under the rubric of theory consists exclusively of recognition for work carried out in simulation, which is quite unrelated to what is usually referred to as fundamental theory.

[5] G. Ozin (1992) Nanochemistry: synthesis in diminishing dimensions, *Advanced Materials*, 4(10): 612–649.

[6] One can point to other gold rushes in the annals of contemporary scientific research. One such case was the high-critical-temperature superconducting materials gold rush. Many scientists took up work in the field, where there was much hope for interesting cognitive development and technical economic yield. In the event, this gold rush has for the moment turned out to be a bubble. H. Nowotny, U. Felt (1997) *After the Breakthrough: The emergence of high temperature superconductivity as a research field*. Cambridge: Cambridge University Press.

The research published by Gimzewski between 1977 and 1983 lay in the sphere of surface physics and related properties. Early efforts dealt with questions and substances such as synchrotron excitation of surfaces and emission of photons, and with ionization energy and anomalous basicity of arsabenzene and phosphabenzene.[7] Other investigations lay in the same surface study domain—exploration of crystalline silicon for its magnetic spin and electron properties. This research dealt, through indirect observations, only with general behavior in the microscopic world; Gimzewski's future endeavors in the nano perspective would instead focus on individual atoms and molecules, their organization in multiple materials and concomitant properties, techniques of how to manipulate molecules, and the technology of scanning probe devices and how they interact with experimental objects. Examination of publications reveals that these projects were sometimes simultaneous or alternatively interspersed, but rarely arranged into stable clusters. This suggests the excitement of early days, and it may also intimate the multiplex, intertwined internal strands of nano as a heterogeneous cognitive and technical entity.

Between 1985 and 1995 (the date at which we close this Gimzewski investigation, which will be extended to the present in Chapter 6), the entirety of the scientist's publications focused on aspects of nano-based issues. The majority of articles investigate the original data that can be gleaned about physical surfaces and properties such as electronic, magnetic, or photon behavior, of a long, long list of substances—copper, gold, silver, selenium, silicon, graphite, etc. In effect, Gimzewski applied the new powerful STM instrument to characterize almost any substance that came to hand in a kind of "shotgun" approach![8] This is perhaps one sign of a youthful science. The task was additionally original because the STM, unlike earlier devices, allowed practitioners to focus on morphology of single objects—a revolution in its own right.[9]

Gimzewski similarly conducted much research on the internal operation of the STM, and more particularly on its sister device the atomic force microscope (AFM), and on how these instruments interact with the objects under study. In 1986 he published a piece reporting an investigation on the forces acting on a sample induced by STM apparatus during measurements. In 1988, two similar articles appeared, one that studied the relation between STM action and photon emission, and another that recommended technical modifications to the STM in order to obtain a more elevated photon yield.[10] The scientist later proposed techniques to enable the STM to probe beneath the surface. Though he conducted less work in this area, Gimzewski's gold rush also entailed investigation of the

[7] D. Fabian, J. Gimzewski, A. Barrie, B. Dev (1977) Excitation of Fe1s core-level photoelectrons with synchrotron radiations, *Journal of Physics F-Metal Physics*, 7(12): 345–348.

[8] J. Gimzewski, A. Humbert, D. Pohl, S. Veprek (1986) Scanning tunneling microscopy of nanocrystalline silicon surfaces, *Surface Sciences*, 168(1–3): 795–800.

[9] J. Gimzewski, E. Stoll, R.P. Schlittler (1987) Scanning tunneling microscopy of individual molecules of copper phthalocyanine adsorbed on polycrystalline silver surfaces, *Surface Sciences*, 181(1–2): 267–277.

[10] J. Gimzewski, B. Reihl, J.H. Combes, R. Schlittler (1988) Photon-emission with the scanning tunneling microscope, *Zeitschrift fur Physik B-Condensed Matter*, 72(4): 497–501; J.K. Sass, J. Gimzewski (1988) Proposal for the simulation of electrochemical charge-transfer in the scanning tunneling microscope, *Journal of Electroanalytical Chemistry*, 251(1): 241–245.

dynamics and morphology of single molecules—an objective that became readily available in nano, thanks to the scanning probe microscopy devices. Along the same lines, he typically researched the capability for manipulating (controlling) single molecules and atoms.[11] Gimzewski's involvement in nanoscience was further reinforced after 1992, when he turned his full attention to STM investigation of the recently synthesized nanostructure material, C60 carbon nanotubes. Between 1992 and 1995, the application of nano instrumentation to this new nanosubstance consumed most of Gimzewski's considerable energies. The rapid response to the emerging area of nano by James Gimzewski is emblematic of the sweep and dynamics of the nanoscience gold rush. Note that he received the Feynman Nanotechnology Prize in 1997.

Christian Joachim, a French physicist based at the Centre d'Elaboration de Matériaux et d'Etudes Structurales (CEMES) in Toulouse (France), is specialized in molecular and atomic dynamical systems, which he explored mainly through theoretical calculations. From the beginning of his career, during the 1980s, Joachim's endeavors were oriented toward exploration of the physical, chemical, mechanical, and electronic properties of atomic and molecular systems.[12] This is probably what led him to the topic of the "molecular switch,"[13] which had been little studied at the time, but which has now become an important theme in nanoscale research, where reversible change in a single molecule and its conformational property are of paramount interest. As we have said and as will be seen in the next case presented here, single molecules constitute a central feature in nanoscale research, and it is notable that from the beginning of the gold rush they became one of the main topics.

Joachim's endeavors in this field revolved around simulation, which was fundamental to his early work on the correlation between single molecules, their switching properties, the STM tip, and the tunneling effect.[14] Here one observes the centrality of simulation work in the gold rush to nano. Observing and imaging at the nanoscale entailed questions and reflections on the physical properties at the molecular scale and of single atoms. Probe microscopy instruments like the STM and the AFM were in themselves the subject of research. As a matter of fact, Joachim (along with James Gimzewski and Reto Schlittler, both at IBM at the time) was awarded in 1997 the Feynman Nanotechnology Prize for his work using scanning probe microscopes to manipulate molecules and for their imaging. As stated by the Foresight Institute, which awards the Feynman Nanotechnology Prize, "in this research, a key element in Dr. Joachim's work has been

[11] R. Berndt, J. Gimzewski (1993) Photon-emission in scanning-tunneling-microscopy—Interpretation of photon maps of metallic systems, *Physical Review B*, 48(7): 4745–4754.

[12] C. Joachim, J.P. Launay (1984) The possibility of signal molecular processing, *Nouveau Journal de Chimie—New Journal of Chemistry*, 8(12): 723–728.

[13] C. Joachim, J.P. Launay (1986) Bloch effective Hamiltonian for the possibility of molecular switching in the ruthenium bipyridylbutadiene ruthenium system, *Chemical Physics*, 109(1): 93–99; C. Joachim (1987) Control of the quantum path target state distance—Bistable-like characteristic in a small tight-binding system, *Journal of Physics A—Mathematical and General*, 20(17): L 1149–L 1155; C. Joachim (1988) Molecular switch—A tight-binding approach, *Journal of Molecular Electronics*, 4(2): 125–136.

[14] A. Aviram, C. Joachim, M. Pomerantz (1988) Evidence of switching and rectification by a single molecule effected with a scanning tunneling microscope, *Chemical Physics Letters*, 146(6): 490–495. It is to be noted that Aviram is one of the first to have worked on molecular switching.

his introduction of elastic scattering quantum chemistry (ESQC) theory to explain tunneling junctions between metal electrodes and molecules, now a standard for STM image calculations," thus emphasizing, the central role of simulation in this episode.[15] In 2005 Joachim was awarded the Feynman Prize for a second time for "developing theoretical tools and establishing the principles for design of a wide variety of single molecular functional nanomachines."

In Joachim's case, one can simultaneously observe a continuity and a break in his scientific trajectory. His interests in calculation-based questions about molecular systems preceded and prepared his total commitment to nanoscale research from the early years of the gold rush. This is visible through his increasingly extensive combination of theoretical and experimental work. It is likewise perceptible in his research on, for example, molecular conformational changes, and dynamical and reversible molecular processes whose concrete development takes the form of single-molecule devices ranging from molecular wires to switches to logic gates to wheelbarrows.

The centrality of the single molecule in nanoscale research, which constitutes one of the vital pivots of Joachim's trajectory and of his work in nano, is also one of the structuring themes of the French physicist Gérald Dujardin's research. Dujardin presently heads the Groupe Nanosciences Moléculaires (which is part of the Institut des Sciences Moléculaires at the University of Orsay, France), whose ambition is to conceive and build architectures at the atomic and molecular scale on surfaces that could be, in the end, susceptible to functioning as nanomachines. Dujardin began his career during the early 1980s in molecular physics in the gas phase, where he specialized in the photo-ionization of molecules detected by spectroscopy.[16] Electron energy, mechanisms of desorption of a surface, and excitation resonance figure among his main topics of study[17] during the period that extends until 1991–2. At that time, he was struck when reading Eigler's article relating his achievement of "writing" the I.B.M. logo with xenon atoms (see Chapter 1), "This has been the most extraordinary event in my career," he reports.[18] It was in this period that Dujardin caught the nano gold rush fever. For him, the possibility to "seeing" individual atoms and molecules opened a completely new perspective for his own research. While in the past he had worked on millions of molecules, therefore on abstract things, molecules were now individual objects that one could see and manipulate. At first, the radical change was not in the questions asked or in the research topics he was pursuing, but in the possibility of controlling and manipulating single molecule shapes and

[15] Simulation models are informed by metrological experiments and the latter rely in part on simulation models. The relationship between simulation and metrology is one of circularity.

[16] For example: G. Dujardin, S. Leach, O. Dutuit, P.M. Guyon, M. Richardviard (1984) Double photoionization of So2 and fragmentation spectroscopy of So2 ++ studied by a photoion photoion coincidence method, *Chemical Physics*, 88(3): 339–353.

[17] For example: Y. Baba, G. Dujardin, P. Feulner, D. Menzel (1991) Formation and dynamics of exciton pairs in solid argon probed by electron-stimulated ion desorption, *Physical Review Letters*, 66(25): 3269–3272.

[18] Interview of Gérald Dujardin (ISM, University of Orsay) by authors, 20 May 2008 and 15 September 2008.

electronic properties.[19] Dujardin and his team progressively learned to control ever more parameters: the tip's electrons' flux, the molecule's geometry, its environment (such as temperature), and its relations with the surface on which it is adsorbed.[20] Dujardin gradually focused on two key facets of the single-molecule nano problematic: the configuration changes in one molecule,[21] and the properties of the surface on which the molecule is located—for example, the question of reconstruction and rearrangement of a surface.[22]

While in the beginning of Dujardin's turn from his first domain of research was perceived by him as a simple transposition of his previous work from large populations of molecules to one individual molecule, his main question evolved. His central research topic now became how to give single molecules a functionality (for example the switching property), and how to build and control nanomachines. In a way, Dujardin's rush into the nano world changed his research paradigm.

The major discontinuity in the research trajectory of Phaedon Avouris, that involved him in the NSR gold rush, is directly connected to adoption of the STM in 1988 and research projects entirely based on the synthesis and exploration of the nanostructured material C60 carbon nanotubes, beginning in 1998. The opening years of Avouris' career had focused on laser spectroscopy and on classical quantum surface properties—radically distant from the nano perspective. Today, Avouris is located at the Tomas J. Watson Research Center of IBM in Yorktown Heights, New York, where he directs nano-related investigations. He also holds adjunct professorships at several US universities, including the University of California in Los Angeles. For Avouris then, the introduction of nanostructured materials and their accompanying nanoproperties revolutionized his cognitive trajectory.

During the late 1970s and early 1980s he published papers on the effects of laser spectroscopy on chemical reactions linked to photo-emission, and photophysical dynamics of aromatics adsorbed on a clean AG (111) surface.[23] In 1982 he turned to a different theme: namely multi-instrument investigations of physical and chemical in situ surfaces' structure and behavior. He set out to explore the surface geometry of adsorption[24] and surface energy loss.

[19] G. Dujardin, R.E. Walkup, P. Avouris (1992) Dissociation of individual molecules with electrons from the tip of a scanning tunneling microscope, *Science*, 255(5049): 1232–1235.

[20] P. Molinas-Mata, A.J. Mayne, G. Dujardin (1998) Manipulation and dynamics at the atomic scale: A dual use of the scanning tunneling microscopy, *Physical Review Letters*, 80(14): 3101–3104.

[21] M. Lastapis, M. Martin, D. Riedel, L. Hellner, G. Comtet, G. Dujardin (2005) Picometer-scale electronic control of molecular dynamics inside a single molecule, *Science*, 308(5724): 1000–1003.

[22] R. Semond, P. Soukiassian, A. Mayne, G. Dujardin, L. Douillard, C. Jaussaud (1996) Atomic structure of the beta-SiC(100)-(3x2) surface, *Physical Review Letters*, 77(10): 2013–2016; P. Soukiassian, F. Semond, L. Douillard, A. Mayne, G. Dujardin, L. Pizzagalli, C. Joachim, C (1997) Direct observation of a beta-SiC(100)-c(4x2) surface reconstruction, *Physical Review Letters*, 78(5): 907–910.

[23] P. Avouris, D.E. Demuth (1981) Spectroscopy and photophysical dynamics of aromatics adsorbed on a clean Ag(111) surface, *Journal of Photochemistry*, 17(1–2): 111–112.

[24] Adsorption is the adhesion of atoms, ions, or molecules from a gas, liquid, or dissolved solid to a surface. This process differs from absorption, in which a fluid (the absorbate) permeates or is dissolved by a liquid or solid (the absorbent). Note that adsorption is a surface-based process, while absorption involves the whole volume of the material <http://en.wikipedia.org/wiki/Adsorption>.

The break occurred in 1988. Avouris continued surface science studies, but this time through experimental investigations using the STM. Recall that the STM was developed in 1981 and won the Nobel Prize in 1986, and within only two years it had already become the centerpiece of his projects. But exactly what advantage did Avouris reap using the STM that went beyond the findings he had obtained with alternative devices? Now equipped with a STM, he could identify individual atoms and their spatial relations, he could observe the surface morphology and internal structure of single molecules, and he could control the position and architectural locations of both in order to generate novel physical properties. The STM's atom-by-atom and molecule-by-molecule grasp of surface position and forces was a far cry from Avouris' former world of microscopic surface science, where information was often only general and based on a more collective understanding. Scores of publications along this line of enquiry punctuated the following ten years, the most novel exploring electronic surface properties.[25] Such work had already marked Avouris' full engagement with NSR, and was followed by a second layer of nano-rooted research which began in 1998.

The year of his switch to carbon nanotubes (1998) as the material basis of his research, as opposed to his previous work on non-nanostructured substances, brought Avouris a rich harvest of important findings. He was among the first to determine in detail key electronic properties of C60.[26] Over the following decade, the nanoscientist concentrated all of his efforts on a combinatorial of the STM and species of fullerenes, and he is today studying the optical and optoelectronic activities of a newly synthesized group of C60. Avouris' path in nanoresearch represents a twin intertwined gold rush—the first in the 1980s, linked to scanning probe microscopy, and the second in the 1990s, connected with the synthesis and exploration of nanostructured materials. Such radical breaks are not uncommon in NSR, yet, as we will now show, variations on gold rush trajectories also occur.

Finally, the research trajectory of Alex Zettl constitutes a variant on the nano gold rush theme. His present work in nano was long preceded by involvement in numerous different projects. The physicist's ultimate nanoresearch orientation is the accumulation of earlier work; he came to it step by step. The pattern is not that of discontinuity and spontaneous rallying. In this respect, it contrasts with the paths of Gimzewski and Avouris. Zettl is today Director of the Condensed Matter Physics Department at the University of California at Berkeley. He is specialized in nanostructures and their properties, and in the synthesis of nanomaterials. By what circuitous path has Zettl ultimately come to nano?

Zettl is a mathematical physicist. He began his career in 1983 with research on semiconductors, studying electronic properties associated with charge–density wave transition

[25] P. Avouris, R. Wolkow (1989) Scanning tunneling microscopy of insulators—Caf2 epitaxy on Si(111), *Applied Physics Letters*, 55(4): 1074–1076; P. Avouris (1990) Atom-resolved surface-chemistry using the scanning tunneling microscope, *Journal of Physical Chemistry*, 95(6): 2246–2256.

[26] P. Avouris, T. Hertel, R. Martel (1998) Manipulation of individual carbon nanotubes and their interaction with surfaces, *Journal of Physical Chemistry B*, 102(6): 910–915.

and non-linear conductivity in Nb-Se3 semiconducting crystals.[27] Zettl continued along these lines for about ten years, however accompanied by significant changes. In 1988 he discovered the STM and began to compare his mathematical results with the physical information offered by the new instrument.[28] In parallel with this, Zettl directed his attention to "single crystals," a signature of NSR, although not unique to it. The introduction of his investigations, in which single crystals comprised a key referent, was tied to the capacity of the STM to observe the relative positions and detailed morphology of individual objects. Although only intermittently, the STM subsequently played a role in parts of Zettl's research—a low-level presence of the nano domain.

Often, with reference to single crystals and the STM, between 1989 and 1991 Zettl focused on high-critical-temperature physics. But this field quickly lost its luster for want of interesting results. During the superconducting years, Zettl also began to study the synthesis of metal materials, and this would indirectly turn him toward nano. The gold rush began for Zettl with the development of C60 carbon nanotubes. Since 1994 the quasi-totality of his research has involved nanostructured materials, where he systematically investigates property after property—electric conductance and resistance, thermal behavior, and optical features.[29] Indeed, reference to nanostructured materials has become ubiquitous in all his publications—the very heart of his efforts. Moreover, consistent with nano, parts of his work revolve around qualitative features. Lastly, Zettl's research center is committed to the development of innovative methods for the synthesis of novel materials, and notably nanotubes. Although initially only gradually moving toward nano, Zettl has finally definitely exhibited a variant gold rush dynamic.

Other elements apart from the progressive or more sudden switch of scientists toward nanoscale questions have played a role in the gold rush that we have so far described through the trajectory of practitioners. What we could call an institutional expression of the gold rush has also contributed to the enthusiasm that has propelled researchers toward the field. Among these, one can cite the Foresight Institute, founded in 1986 by Eric K. Drexler, a militant of molecular nanotechnology. The Feynman Nanotechnology Prize, established in 1993, is a child of this Institute. Grassroots movements have similarly contributed to the rush. The enthusiasm for nano, spawned in the early 1990s in favor of expending research on the synthesis of nanostructured materials, as provoked by the Canadian nanochemist Geoffrey Ozin, illustrates the impact of collective efforts on the NSR gold rush.

Drexler is an American engineer who is frequently viewed as the prophet of future NSR; his early efforts provided one of the first organized cognitive and institutional

[27] A. Zettl, G. Grunier (1983) Charge-density-wave transport in orthorhombic Tas3 0.3. Narrow-band noise, *Physical Review B*, 28(4): 2091–2103.

[28] R.E. Thomson, U. Walter, E. Ganz, J. Clarke, A. Zettl, P. Rauch, F.J. Disalvo (1988) Local charge-density-wave structure in 1t-Tas2 determined by scanning tunneling microscopy, *Physical Review B*, 38(15): 10734–10743.

[29] For example: J.G. Hou, X.D. Xiang, V.H. Crespi, M.L. Cohen, A. Zettl (1994) Magnetotransport in single-crystal Rb3C60, *Physica C*, 228(1–2): 175–180; J.G. Hou, X.D. Xiang, M. Cohen, A. Zettl (1994) Granularity and upper critical fields in K3C60, *Physica C*, 232(1–2): 22–26.

pushes in NSR.[30] During his studies, in 1979, he was struck by Feynman's provocative 1959 talk, "There is plenty of room at the bottom," and this drove him to invest in what he later called "molecular nanotechnology." In 1986, together with Christine Peterson, he founded the Foresight Institute "to guide emerging technologies to improve the human condition." His main interest here was to focus efforts on "nanotechnology, the coming ability to build materials and products with atomic precision."[31] The Institute rapidly acquired an audience, and in 1993 it created the Feynman Prize, which rewards researchers whose work has most advanced the development of molecular nanotechnology. The Foresight Institute and its Feynman Prize have contributed to the nano gold rush in three important ways. Since early days, long before nano was on the official agenda, the Institute promoted both the concept and concrete research. It made nano known when few people had heard of the word. On another register, it awarded highly visible prizes for outstanding research results. Nanoscience and scientists became increasingly visible. Nano was acquiring its *lettres de noblesse*. Finally, and perhaps of foremost importance, by selecting specific domains of research for prizes, the Foresight Institute has progressively affected the themes and analytic tools of NSR. It has decidedly oriented the emergence and evolution of research topics. The Institute and the Prize have notably encouraged research in four directions: (1) biology-related work, (2) the themes of single molecules, (3) control and switching, and (4) metrological and simulation instrumentation. The laureates of the Prize have frequently emerged as leaders in the diversity of domains that comprise NSR.

The nano gold rush also benefited from grassroots movements. The efforts of Geoffrey A. Ozin illustrate that dynamics. In 1992 this nanochemist published what immediately became a classic article on innovative nanomaterials in the landmark journal, *Advanced Materials*. For over a decade "Nanochemistry: Synthesis in diminishing dimensions" remained the journal's most cited article.[32] It proved highly influential, as attested by the fact that it had been cited over one thousand times by 2012. Ozin, perhaps more than any other single nanochemist, became identified with the growth of the field.

Geoffrey A. Ozin studied at King's College, London and Oriel College, Oxford University. He is now Professor of Chemistry at the University of Toronto and a Founding Fellow of the Nanoscience Team at the Canadian Institute for Advanced Research. He is considered to be one of the fathers of nanochemistry. His research includes studies of new classes of nanomaterials, photonic crystals and, most recently, nanomachines. In fact it is not solely the idea, the technicity, and the orientation toward mastering and creating nanomaterials in a very controlled way that Ozin was introducing in his article to the chemistry and materials science communities. It was also the crucial position of these domains of research, which now had the tools and the research horizons to become central in the nanoscience communities, because they could provide laboratories in physics,

[30] H. Choi and C. Mody (2013) From Materials Science to Nanotechnology: Institutions, Communities, and Disciplines at Cornell University, 1960–2000, *Historical Studies in the Natural Sciences*, 43(2): 121–161.

[31] <http://en.wikipedia.org/wiki/Foresight_Institute>.

[32] G. Ozin (1992) Nanochemistry: Synthesis in diminishing dimensions.

in chemistry, and in biology with these nano-tailored materials. In all of his efforts, Ozin clearly capitalized on the 1992 radical-material's revolution, sparked by Iijima's synthesis of C60 carbon nanotubes (see Chapter 1); and with his characteristic enthusiasm, Ozin extended the gold rush a step beyond.[33]

2.2 MATERIALS, RESEARCH-OBJECTS, AND PHYSICAL PHENOMENA

In this section of the chapter, we focus on what we term "research-objects," and more specifically on their particularities in the NSR synergistic system of cognition. What is intended by the term research-object? We draw attention to a crucial distinction between materials and research-objects. For our purposes, in NSR, a material is a large family of artificial substances that are in general used in the composition of downstream commercial or research components. There exist many families of materials, such as fullerenes or semiconductors. The synthesis, production, and use of a material entail considerable specialized research, the development of fabrication routines and product standardization. By research-objects we mean specifically tailored substances (sometimes one of a kind), derived from one or several families of nanostructured materials, intended to be a terrain for the exploration of particular physical properties. Research-objects are designed by metrological experimenters in order to address often well-defined questions in the context of a particular project and even a specific experiment. The work of materializing the research-object demanded by experimenters is often undertaken by nanopractitioners referred to as "epitaxiors" (see Chapter 1), whose specialty consists of building tailored substances. One example of a research-object that has been designed for a specific

[33] In the preceding pages, growing enthusiasm for nanoscale research in the physical sciences has been discussed, principally in terms of the capacity of the STM, and later the AFM, to observe aspects of previously undetectable phenomena, or to study them with enhanced precision. In combination with this, new families of artificial materials were generated, particularly as low-dimension substances. This work may be seen as representing the power of curiosity in research. In an article by M. Lynch and C. Mody, the authors have suggested that curiosity was particularly acute in the growing field of surface science, and specifically regarding the solution to a thorny long-standing problem. The STM allowed practitioners to effectively explore the surface reconstitution of a highly complex crystal – the 7x7 for silicon (111). Lynch and Mody refer to this crystal as a "test-object," which was emblematic of effective research that allowed the calibration of instruments and later on served as a pedagogical platform (C.C. Mody, M. Lynch, 2010) Test objects and other epistemic things: a history of a nanoscale object, *The British Journal of The History of Science*, 43(03): 423–456). On a quite different register, J. Hennig proposed that the way in which images are designed when using the STM contributed significantly to the success of this device. In the first ten years of the STM's operation, the architecture of images changed at least five times. Ultimately, they became topographical representations of nanoscopic objects where the scale is stipulated and color is used restrictively (J. Hennig, 2004). Changes in the design of scanning tunneling microscopic images from 1980 to 1990, *Techné: Research in Philosophy and Technology*, 8(2). Some scientists move beyond this style of representation where color becomes primary and connections between topology and rendering of objects is approximate (T.W. Staley, 2008) The coding of technical images of nanospace: Analogy, disanalogy, and the asymmetry of worlds, *Techné: Research in Philosophy and Technology*, 12(1): 1–22).

research project and developed from the generic III–V family of nanostructured thin-film semiconductors is a semiconductor quantum dot composed of gallium and arsenic, located in a specially configured aluminum substrate used for research on optical radiative and propagation properties. Here the original synthesized material is the generic zero-dimensional quantum dot that has been tailored to satisfy a specific research goal.

Here we explore the centrality of research-objects in metrological experimentation and analyze their sometimes complex and precarious position located between epitaxiors, who design and make them, and metrology-based experimenters, who insist on designing them. Research-objects thus lie between experimenters and epitaxiors, and they are sometimes shared by them and at other times a terrain of dispute lies between them. It is not infrequent that stubborn silence persists between epitaxiors, who are themselves acknowledged specialists having their own established market, and experimenters. But dialog can also be abundant. Such dialog is seen in many quarters as essential because it allows elasticity in the process of research. As scientists gain understanding in the course of their experiments, they request modifications from epitaxiors, and the resulting new research-object yields an upward spiral of knowledge.

Conversely, the producers of research-objects may themselves modify their materials, and with this they sometimes convince experimenters to redirect their questions or their experimental set-up in order to explore the new object. This constitutes a dynamical process of cognition/object dialog that is one methodological signature of NSR. This dynamic will now be described for two research teams: one that explores photoluminescent emission and propagation in quantum dots, and the other which works on optical and acoustical wave propagation in nanocavities.

2.2.1 The photoluminescence case study

The case we now describe identifies some cognitive, material, and instrumental factors that accompanied the trajectory of a research group from classical optics to nanoscale research in photoluminescence, where relations between epitaxy and experimentation figure centrally. Photoluminescence is a process in which a substance absorbs photons (electromagnetic radiation) and then re-radiates them. The early research of the team director, Roger Grousson, focused on the optical superposing of images for storage purposes and on hologram optics, during the late 1970s and 1980s. Grousson's team is part of the Institut des Nanosciences de Paris, founded in 2005.[34] This institute is in part a reformatting and assembly of previous laboratories, principally the Groupe de Physique des Solides, which has functioned since the 1970s. In 1993 Grousson and his team began to publish abundantly on optical phenomena in nanoscale objects. Three factors made this transformation possible: (1) The invention and availability of picosecond lasers, and intra-laboratory construction of a high-performance wave-guide, permitted the exploration of photon dynamics of a new sort where quantum phenomena can be explored. (2) A large

[34] <http://www.insp.jussieu.fr/>.

well-equipped epitaxy laboratory, the Laboratoire de Photonique et de Nanostructures (LPN Marcoussis-France), specialized in three to five semiconductors was developed which suddenly gave access to low-dimensional materials in the form of the synthesis of three to five substances; the objects studied by the team were quantum wells (two-dimensional substances), which had existed for almost two decades but were not easy to acquire locally. (3) Finally, the recent metrological and analytic centrality of excitons as a topic of study in photoluminescent processes became increasingly powerful topics of research.[35] The combination of these three elements spelled the way to nanoscale research for Roger Grousson's team.[36] Viewed retrospectively, the study of photoluminescence has indeed constituted an important research focus in single nano-objects. In this description of work, the term "single nano-object" refers to the investigation of a single crystal, be it in the form of a 2-d nanowell, a 1-d nanowire, or a 0-d quantum dot. This capacity to conduct research on a single nanostructured crystal distinguishes NSR from most previous science. The "single nano-object" is a physical entity and beyond this it represents a pivotal concept of nanoscale research-object at large. In the first stage of their nano investigations (1993–6), the team improved its technique for the study of single nano-objects (here quantum wells) in the framework of the examination of photon excitation (absorption index) and emission, and taking into consideration the limits of nanowells and possibilities of alternative materials (nanowires).[37] Nanowires are one-dimensional materials. They began to become objects of research in 1993 with five publications, rising to 26 publications in 1995 and to 94 in 1997. The team published its first article that year and thus count among the pioneers of this research area. The decision to move from wells (two-dimensional material) to wires (one-dimensional material) was related to the investigation of single nano-objects under certain technical conditions, as in the case of wave-guides. In addition, wires also yield more precise information on excitons. As just indicated, at this time nanowires were a novel rarity and quite difficult for nano experimenter physicists to acquire. The local network of epitaxy acquisition that had provided the quantum wells did not synthesize the required sort of wires. The desired wire had to be made of a particular material (GaAs and AlAs), having a specific shape needed for the prospective research. The shape of the wire took the form of a V, and this configuration has subsequently had an unintended impact on future research. But for the time being, experimenters lacked a supply network. A recently created large, well-equipped,

[35] An exciton is a quasi-particle found in semiconducting materials. It is a pair formed by an excited electron which has acquired a higher energy level and the resulting "hole." When the excited electron recombines with its hole, energy is generated which takes the form of a photon emission. See L. Apker, E. Taft (1950) Evidence for exciton-induced photoelectric emission from F-centers in alkali halides, *Science*, 112(2911): 421–421.

[36] Interviews with Roger Grousson and Valia Voliotis, and Marco Ravaro by authors between March 2008 and March 2009 at the INSP.

[37] V. Voliotis, R. Grousson, P. Lavallard, E.L. Ivchenko, A.A. Kiselev, R. Planel (1993) Gamma-x mixing in type-ii GaAs / Alas short-period superlattices, *Journal de Physique IV*, 3(5): 237–240; V. Voliotis, R. Grousson, P. Lavallard, E.L. Ivchenko, A.A. Kiselev, R. Planel (1994) Absorption-coefficient in type-ii GaAs / Alas short-period superlattices, *Physical Review B*, 49(4): 2576–2584.

and highly competent national epitaxy center (the Laboratoire de Photonique et de Nanostructures), headed by a colleague, could in fact have synthesized the wanted sample. Grousson's team was indeed associated with this epitaxy network. The fact that the epitaxy lab failed to respond positively to this request shows the extreme specialization of epitaxy work, and the huge inertia that governs many activities.

At that point, the team undertook a literature search for articles on nanowires. They discovered only three epitaxy groups that could provide the specific type of nanowire they needed for their single nano-object photon luminescence work: one team in Switzerland, one in England, and one in Japan. From the Swiss, they received no reply; the English epitaxy team refused to provide wires because it was conducting its own experiments; the Japanese sent samples within a week (the Electron Devices Division, Electrotechnical Laboratory, Tsukuba Japan). This episode exemplifies the dependence relations between experimenters and epitaxiors. Had there been no answer from any epitaxior, the project could well have foundered. This is not to suggest that nanoscale research is dictated by epitaxy, but it does demonstrate that epitaxy constitutes a serious limiting condition. In the case of Grousson's team and the Japanese, their initial dealing evolved into a long-lasting interaction. This Japanese team is the provider of nanowires for other groups, like the Laboratory of Theoretical Physics of Nanosystems in the Ecole Polytechnique in Lausanne (Switzerland).[38] The Japanese team is also engaged in transistor-related epitaxy.[39]

The sample provided by the Japanese epitaxiors, Wang and Ogura, was made by etching V-shaped parallel lines into an aluminum (Al) flat substrate and then filling the bottom of the V with a gallium arsenic nanowire (GaAs). Exploring a single nanowire, Grousson's team could accelerate its study of electro-photon luminescence dynamics.[40] The team soon observed the presence of defects in the nanowire, and in a discussion with the epitaxiors, Wang and Ogura's team, it was discovered that these defects were in effect quantum dots.[41] The observed irregularities corresponded with different light wavelengths, which are markers of quantum dots (see Chapter 1). Grousson then requested that the Japanese team produce quantum dots of different tailored sizes encapsulated in V-shaped nanowires—an instantiation of "research-objects," or otherwise stated, "objects by design." As shown in Figure 2.1, the highly sophisticated laser excitation system developed by Grousson allowed the team to stimulate a single nanodot in a cluster of dots which possessed a different signature because of their different size. This yields

[38] A. Feltrin, K.R. Idrissi, A. Crottini, M.A. Dupertuis, J.L. Staehli, B. Deveaud, V. Savona, X.L. Wang, M. Ogura (2004) Exciton relaxation and level repulsion in GaAs / AlxGa1-xAs quantum wires, *Physical Review B*, 69(20).

[39] C.K. Hahn, T. Sugaya, K.Y. Jang, X.L. Wang, M. Ogura (2003) Electron transport properties in a GaAs / AlGaAs quantum wire grown on V-grooved GaAs substrate by metalorganic vapor phase epitaxy, *Japanese Journal of Applied Physics Part 1–Regular Papers Short Notes & Review Papers*, 42(4b): 2399–2403.

[40] J. Bellessa, R. Grousson, V. Voliotis, X.L. Wang, M. Ogura, H. Matsuhata (1997) High spatial resolution spectroscopy of a single V-shaped quantum wire, *Applied Physics Letters*, 71(17): 2481–2483.

[41] J. Bellessa, V. Voliotis, X.L. Wang, M. Ogura, H. Matsuhata, et al. (1997) Evidence for exciton localization in V-shaped quantum wires, *Physica Status Solidi A-Applied Research*, 164(1): 273–276.

Figure 2.1 The instrument set-up of the optical spectroscopy experiment. The nanostructured semiconductor sample lies inside a liquid helium cryostat. The sample is excited by a laser and the emitted light is detected and analyzed in order to explore the electronic properties of the nanostructured object. The image shows mirrors, lenses, polarizers, beam splitters, and in the center a cryostat. Image provided by Valia Voliotis and colleagues, reproduced with permission of the INSP.

exceptionally rich information on processes of photon emission (see the schematic of photoluminescence in Figure 2.2). One can count over 12,300 published articles dealing with quantum dots and photoluminescence between 1997 and 2012.

For some, the environment (the aluminum substrate) of the nanowires and the quantum dots became a subject for study. Here the question was to identify the quantity of luminosity issuing from the quantum wire or dot, and the amount of luminosity associated with the shape of the substrate. First of all, epitaxiors displaced the nanowire from the bottom of the V to its wall, and secondly they developed samples with different spacing widths between the apex of the Vs. Some samples had a relatively broad horizontal surface between the apex of two Vs, some others had such a small bridge that two adjoining Vs formed a W shape. Wang came to Paris with his samples in order to study the effect

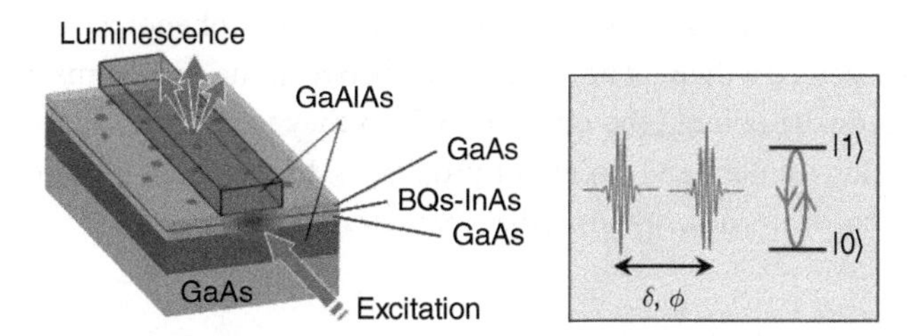

Figure 2.2 Schematic of photoluminescence. Image reproduced with permission of Valia Voliotis.

of these modifications on light propagation with Grousson's team's specialized instrument set-up. In important respects, interactions between experimenters and epitaxiors constitute a kind of collaboration. This work led the experimenters to focus more definitely and clearly on the contribution to luminosity generated by photon excitation and environment-produced effects. By separating these two components, it became possible to identify the precise consequences of strictly photon-based dynamics. This reinforced the part of Grousson's team associated with propagation. For the Japanese epitaxiors, it provided useful information about substrates and active low-dimensional objects, and thus enhanced the capability of control over the potential relations between a tailored object and the effects that one can generate using it.[42]

In this episode, one can observe four types of interaction between epitaxiors and metrology experimenters. The first category may be termed "demand/supply." Grousson's team initiated a search for a highly specific sample, tailored to correspond with a particular experimentation project; and the Japanese supplied an object they had already developed. Interaction 2 involved a surprising observation of the existing sample and a request to the epitaxy team for it to adjust its object accordingly. This constitutes an "observation/adjustment" relationship. This second category of interaction is a balanced interaction, as opposed to the first category, which is experimenter driven. The third category of interaction is one in which a prolonged normalized exchange of suggestions and reactions on both sides routinely occurs. One observes a flow between experimenters and epitaxiors. Contrary to the first two exchanges, this is a stable, long-term collaboration. In the fourth interaction, experimenters and epitaxiors became interested, for different reasons, in the impact of the nanomaterial environment on physical effects. The nanophysicists' interests lay in identifying and understanding the contributions of exciton decay and electromagnetic features to the luminosity of their system. The Japanese epitaxiors were interested in developing mastery of their product. This reflects this team's occasional involvement in the advance of technological applications, such as transistors.

These different types of interaction are emblematic of nanoscale research for at least two reasons. (1) As explained in Chapter 1, NSR is a combinatorial science. Combinatorials can, as in this case, rely on the high level of technical skill—here epitaxy on the one hand, and on the other, the sophisticated experimental set of devices and procedures of Grousson's team which make scientists of the two groups dependent on each other. (2) As will be seen throughout the chapters of this book, NSR is a descriptive science. Here, as in the majority of projects, the idea is to master a series of phenomena in order to better observe and describe them. The interactions between the two teams of the episode related here gradually tailored the object on which they were working in order to generate an increasing number and variety of properties. In this sense, this example reveals important aspects of the descriptivism that characterizes NSR.

[42] X.L. Wang, S. Furue, M. Ogura, V. Voliotis, M. Ravaro, A. Enderlin, R. Grousson (2009) Ultrahigh spontaneous emission extraction efficiency induced by evanescent wave coupling, *Applied Physics Letters*, 94(9).

2.2.2 The nanocavity case study

We will now present a second and even more complex instance of interaction between metrological experimentation and epitaxy. In this episode, a triangular relationship develops. Two research specialties, one in optics and the other in acoustics, are involved where epitaxy serves the independent requirements of each, and at the same time, constitutes the junction that meaningfully links the two specialties to each other. The link is such that both research domains benefit individually and in common. In the one instance, the development of a new research-object permits the transformation of a theoretical perspective into a concrete research project.[43] In the other instance, the introduction of the shared object allowed the concrete generation of a new range of physical phenomena that had long been blocked by the absence of the appropriate substance.

Bernard Jusserand is a physicist specialized in optics, presently working at the Institut des Nanosciences de Paris (INSP). He predominantly uses Raman light-scattering techniques to explore photons and phonon propagation (half of his 162 publications are associated with Raman spectroscopy). His research has focused on a variety of nanomaterials, such as quantum wells, nanowires, and micro- and nanocavities. In line with this, he has developed a concern for epitaxy techniques—their possibilities and limits. He has acquired considerable knowledge in this domain. One of his doctoral students, Aristide Lemaître, working at the Laboratoire de Photonique et de Nanostructures (LPN), was to become a specialist in epitaxy, and as will be seen below, came to play a pivotal role in this story. Compared with most other practitioners in nanoscale research that we investigated in our study, Jusserand's profile is particularly rich because it extends to such a large number of materials, research questions, and diversified but connected domains of investigation. In particular, it seems that few nanophysicists who work in optics exhibit at the same time such a sustained interest in phonons. Between the late 1980s and *c.*2000, Jusserand explored some of the physical properties of these materials. Raman spectroscopy also allowed him to generate sonic waves at extremely high frequencies—terahertz. In contrast with this, over this period, the technique of sound pumping was limited to gigahertz. Attainment of ultra-high frequency could only be achieved in relatively low-complexity conditions, which limited the richness of results. In this situation, the possibility of constructing analogies between acoustical waves and optical waves was compromised. Jusserand consequently oscillated between pursuing or abandoning this important topic. He nevertheless remained interested in the subject, but was unable to see how to advance in these investigations.

The second physicist relevant to this case study is Bernard Perrin, who is specialized in acoustics. His research focuses on attenuation of acoustical waves in metallic superlattices. In 2002, in the context of a scientific meeting, he discussed his research with

[43] Interviews of Bernard Perrin, Bernard Jusserand (INSP, University of Paris) by authors, 22 October 2008, 3 March 2009; interview of Florentina Pascual-Winter (INSP-University of Paris) by authors 4 December 2008; interview of Agnès Huyn (INSP, University of Paris) by authors, 12 May 2008.

Jusserand, alluding to the study of elastic properties of extra-thin materials and the need to generate ultra-high frequencies for the exploration of fundamental problems in physics, and notably in his domain of the propagation of acoustic waves. In well-organized materials like crystals, waves propagate well. In contrast to this, in less well-structured materials, above a certain frequency, they exhibit localized behavior. He then wanted to examine the gap between wave propagation and localized waves. At the time, being limited to gigahertz by his instruments, a pump-probe pico laser, and by the materials, Perrin was unable to generate the terahertz waves that were essential to his experiments. For this, he desperately required a new research-material. In the same way that Jusserand was interested in the connection between sound and light, Perrin's attention lay in the relations between sound and heat; the latter exhibiting the same nature of vibrations as sound. The complexity of their common project constituted simultaneous stimulus for further development in their own particular spheres, including heat and theory on wave propagation for Perrin and, as indicated above, convergence between light and sound for Jusserand. Perrin's research problem drew Jusserand's attention. They met and perceived that they could collaborate on their shared interest in vibration phenomena. Each scientist confronted his specific research problem, but found sufficient terrain to find advantage in working together. Perrin's major difficulty lay in the lack of access to research-objects tailored to his needs.

There were two problems: Perrin was using material samples that had been created for magnetism-oriented research, for example, the manufacture of reading heads, and which were inappropriate for his acoustical studies. The second problem was the silence, or rejection that he encountered when he asked for samples from epitaxy groups like the Institut d'Electronique Fondamentale (Strasbourg) and the Laboratoire d'Electronique du Poitou. Up to this point, Perrin had worked with metal-based samples, which he saw as blocking his research: in order to advance, he perceived a necessity for three to five semiconductor materials. Perrin's difficulty demonstrates the requirement to be part of an epitaxy network, to be able to identify people who share the same interest or to locate teams that have already synthesized precisely those materials being requested. In the absence of properly designed and synthesized materials, NSR is quite simply unthinkable. Materials-by-design are an insurmountable characteristic of nanoresearch.

Each of the two experimenters perceived benefits in collaboration: Perrin could get from Jusserand access to research-objects and to his knowledge of these materials, and also to his privileged connection with an epitaxior (his former student). Jusserand, in turn, could become significantly involved in terahertz and their interpretation, to ultimately be able to formulate analogies between acoustical and optical events. In effect this constituted a concrete window of opportunity in a domain that had long fascinated him and to which in the past he had only been able to contribute fitfully. By working with Perrin, Jusserand could investigate aperiodic materials.

The collaboration began with a gift from Jusserand to Perrin of a three to five semiconductor sample that he had in his drawer, suitable for Perrin's exigencies. In a first article, published in 2004, the two scientists used their respective excitation devices, one a Raman and the other a picosecond laser pump-probe, to produce and study sound waves in a micro-cavity. Both Perrin and Jusserand had worked with micro-cavities in their preceding research on super-lattices, and on which Perrin had published articles belonging to this family of objects. Using the same sample, the two scientists could study two kinds of perturbations in a micro-cavity, one with light waves, the other with sound waves.

The perturbation induced by light can be referred to as an impulse. It may be likened to a swift hammer tap on a pendulum, which oscillates and then returns to its equilibrium state. The perturbation induced by sound can be described as a pendulum that is moved from one support to another. In this case, instead of immediately returning to its initial equilibrium, the system gradually finds a new equilibrium state. In the case of the impulsional light perturbation, the time is brief. In the case of the percussional perturbation, the time required to obtain equilibrium is relatively long, ranging from picoseconds to nanoseconds. For the two scientists, working together allowed them to explore the complex relations between the two forms of perturbations. The two-wave systems occurring together generate a family of highly original and complex phenomena.[44] Nevertheless, with the first sample which they began to work on together, they could not obtain sufficiently elevated frequencies (terahertz were not attainable). In order to generate these, a new material would be required. This would take the form of a specially-tailored-to-need nanocavity, which was not then available to them. It thus had to be designed and synthesized. This task would fall to the epitaxior, Aristide Lemaître.

The pre-history of the nanocavity begins in 1992, when it was first predicted and described by simulation-based results;[45] and the first physical nanocavities were synthesized two years later, in 1994.[46] A nanocavity consists of a nanometric layer having specific properties located between two super-lattices. These objects proved extremely difficult to produce, and consequently their introduction into experimentation was slow (publication of only one article on this object in 1995, three in 1998, eight in 2000). It was only in 2005 that one can see the beginning of a stable and significant growth in the number of publications (91 articles). At this point, interest in the topic was not yet evident, and it is only from 2005 onwards that a momentum of concern, both in epitaxy works and in optics physics, really began to be perceptible. Here it is remarkable to note that the two most-cited articles dealing with nanocavities focus on epitaxy.[47]

[44] P. Lacharmoise, A. Fainstein, B. Jusserand, B. Perrin (2004) Semiconductor phonon cavities, *11th International Conference on Phonon Scattering in Condensed Matter, Proceedings*, pp. 2698–2701; A. Huynh, N.D. Lanzillotti-Kimura, B. Jusserand, B. Perrin, A. Fainstein, M.F. Pascual-Winter, E. Peronne, A. Lemaitre (2006) Subterahertz phonon dynamics in acoustic nanocavities, *Physical Review Letters*, 97(11).

[45] Y.C. Lian, J.C.M. Li (1992) A nanocavity in a FCC crystal, *Materials Chemistry and Physics*, 32(1): 87–94.

[46] C. Seager, S.M. Myers, R.A. Anderson, W.L. Warren, D.M. Follstaedt (1994) Electrical-properties of He-implantation-produced nanocavities in silicon, *Physical Review B*, 50(4): 2458–2473.

[47] Y. Akahane, R. Asano, B.S. Song, S. Noda (2003) High-Q photonic nanocavity in a two-dimensional photonic crystal, *Nature*, 425(6961): 944–947; T. Yoshie, A. Scherer, J. Hendrickson, G. Khitrova, H.M. Gibbs, G. Rupper, C. Ell, O.B. Shchekin, D.G. Deppe (2004) Vacuum Rabi splitting with a single quantum dot in a photonic crystal nanocavity, *Nature*, 432(7014): 200–203.

The perceived necessity to obtain a nanocavity demonstrates the centrality of research-objects within nanoresearch. Research-objects determine the research that can be conducted. We will now see how the dialog between the experimenter physicists and the epitaxior who provides them with his tailored samples, structures the step-by-step work of defining and complexifying the experimental research question: the experimenters have questions that the epitaxior "translates" into the configurations and properties he is able to give to the materials; and it is these properties that the experimenters investigate. The evolution of Perrin and Jusserand's subsequent exploration would depend on modifications of samples provided by Lemaître in his epitaxy laboratory.

Lemaître's pivotal role consisted of constructing the nanocavities that were crucial to experiments. The conceptually and experimentally advanced and complex joint project of Perrin and Jusserand called for a highly sophisticated research-object consisting of two major modifications. The nanocavity has two Bragg mirrors,[48] which confine optical and acoustical waves, and amplify them. This set-up allowed Perrin to use the physical object as a novel detection system. In most instances, the surface of the cavity that is excited to produce optical and acoustical effects also serves as a platform for the detection and study of vibration. The inconvenience of this approach is the differentiation of input and output. Perrin used the opposite side of the cavity as an alternative surface for detecting vibration, which yielded high-precision information.

In some of Jusserand and Perrin's research, they insert a nanocavity inside a micro-cavity. This research-object posed acute epitaxial problems for Lemaître. It is no easy task to place a nanocavity inside a micro-cavity. The need to add a second Bragg mirror to a nanocavity meant a rethink of how to tailor this cavity. Nanocavities with Bragg mirrors have one smooth surface and another opposing rough surface. But in this case, the second mirror required a second smooth surface. The challenge consisted of generating the second smooth surface without abrading the first surface. In the discussion of what was required and reflections on how it could be achieved, Lemaître introduced several suggestions about what modifications to the object would be needed for the exploration of particular physical properties—such as the production of elevated terahertz. In some instances, the routine communication between Jusserand and Lemaître was complemented by joint meetings with Perrin, where Lemaître intervened in his capacity as a physicist and not only as epitaxior.

With the complex tailored nanocavity, Jusserand's long-standing theoretical ambition to study the continuity between optical and acoustical waves in terms of phonons more exactingly became experimentally possible. Previously fragmented investigation of each of the two fields could now converge. For his part, Perrin's objective to generate ever higher acoustical frequencies was realized thanks to a novel artifact. By moving from gigahertz to terahertz, he was able to explore a new acoustical vista and also to move toward assimilation of acoustical waves with heat waves. The goal of both scientists in their specific domains was predicated on the design and construction capability of epitaxy

[48] A Bragg mirror is a structure formed from multiple layers of alternating materials with varying refracting index.

in the person of Lemaître. The joint advantage of the collaboration was expressed by Jusserand and Perrin in the following terms:

> The availability of precise technical set ups for the generation and detection of acoustical waves, in the order of terahertz, opens numerous possibilities for spectroscopy, imagery and the physical study of vibrational propagation in solids . . .[49]

Their shared research results are reported in their numerous jointly signed articles—some 24 between 2004 and 2012—in top-rated journals such as *Physical Review B, Applied Physics Letters, Physical Reviews Letters*. They are also often cited in these and other high-ranked journals. Their work has met with considerable success, as illustrated by the birth of a research summer school entitled "Son et Lumière" (held every two years),[50] where Jusserand's guiding concept of an understanding about integrating optical and acoustical vibrations (in terms of phonons) is the central concept, and where Perrin's idea of the continuity of this concept into heat may also be envisaged.

From its very beginning, Jusserand and Perrin's project to work together can be seen as a synergistic dynamics: through their collaboration and the constant participation of Lemaître, Perrin and Jusserand have contributed to opening a field of research which is in itself at the juncture of two different problematics in physics: optical and acoustical wave propagation in a nanoscale research-object. This synergy renders even more relevant the question of control over the generation, description, and study of properties in nanoscale objects, and opens the way to broader horizons in other fields in physics research. Here epitaxy serves to stimulate research in two different research domains and functions as a pivot between them.

Perhaps most significantly, in NSR, the centrality of research-objects to projects brings to light the existence of a new relationship between the research question and the concrete possibility of experiment. In contrast with the past, when scientists often limited their questions to the possibilities of materials existing in nature, now in NSR practitioners are free to formulate questions in the knowledge that, through epitaxy and other materials synthesizing techniques, the desired research-objects carefully tailored to need can probably be generated. Such research by design constitutes a hallmark of NSR.

2.3 METROLOGICAL EXPERIMENTATION AND SIMULATION-BASED EXPERIMENTATION: BETWEEN DIFFICULT ENCOUNTERS AND SYNERGY

The expansion of numerical simulation-driven experiments since the 1990s is particularly dramatic in nanoscale research, where the latter provides an inviting material and

[49] <http://www.insp.upmc.fr/Nanostructures-pour-la,168.html>; <http://www.insp.upmc.fr/-Presentation-.html.>

[50] <http://www.cab.cnea.gov.ar/sonetlumiere/.>

cognitive environment for convergence and intertwining of metrology experimentation and simulation. A newly emergent synergistic tandem between metrology and simulation experimentation is increasingly common, however this relationship does not prevent autonomous projects by the two groups.

These emerging orientations and complementarities are readily traced in the domain of nanoscale studies of surface science. Surface science constitutes a historical and strong theme in twentieth-century research. It began in the early part of the century, and its development equates with the discoveries of the chemists Paul Sabatier (Nobel Prize in chemistry in 1912), Fritz Haber (Nobel Prize in chemistry in 1918), Irving Langmuir (Nobel Prize in chemistry in 1932), and most recently, Gerhard Ertl (Nobel Prize in chemistry in 2007). Much surface physics deals with phenomena of molecular and atomic interfacing—solids and liquids, solids and gases, and gases and liquids. Interfaces are frequently studied with reference to charge, polarization, and energy.

The significance of a nanoscale perspective in surface science is that scientists, simulators, as well as metrologists, can study interfaces and surfaces at the atomic or molecular level (often in terms of single objects or deterministic, stabilized molecular ensembles), where they deal with the same objects and share a mutually intelligible language and set of images. Work processes and cognition of such research can be seen in the activities of a mixed metrology/simulation surface physics team, located at the Institut des Nanosciences de Paris (INSP), which mainly investigates hydroxylation and dissolution, specifically in thin films.[51] This team focuses on oxide surfaces in nano-objects, which means thin oxide films, as opposed to bulk materials. They are particularly interested in the contact between oxide surfaces (frequently magnesium oxide) and water. The project consists of describing and explaining the surface configuration of atoms and their dynamics, and the question of dissolution mechanisms.

The simulators and metrologists studied here exhibit common features and also numerous particularities. As indicated earlier, the power of contemporary simulation resides in the development of algorithms and computational capacities that allow the treatment of an ever-increasing number of atoms. Using high-powered computers, simulators can effectively calculate the position and operation of a thousand atoms, or even thousands of atoms and their interactions. Some simulators declare that this development constitutes a better "fit with reality,"[52] and better reflects the results of experimentation.

The restriction to a small number of atoms had long confined theoretical reflections to limited questions. The ability to include increasing quantities of atoms entails the possibility of dealing with a diversity of elements and their relations, and this complexity has led to the elaboration of reflection on the content and structure of physical systems at the nanoscale. In the confined spaces relevant to nanoscale research, what counts is not so much the number of atoms; it is instead to describe and analyze the complexities of the positions and relations of a spatially circumscribed ensemble, including forces.

[51] Interviews of Claudine Noguera by authors, at the INSP, 23 July 2007.

[52] Interview of Jacek Goniakowski (INSP, University of Paris) by authors, 3 May 2008.

Algorithms are the heart of simulation. The work of many simulators includes writing new algorithms or adapting existing ones. The algorithms used in a particular simulation experiment are adapted to specific research materials and questions. They include certain hypotheses, focal points, and questions expressed through the introduction of numerical values and hierarchies of relations. As new materials are constantly being synthesized and additional research-objects are designed for study in nanoscale research, novel algorithms are correspondingly prepared. Algorithms express idealized substances and forces. Idealized substances and forces are viewed by simulators as "approximations." The approximations are grounded on quantum theory, semi-empirical information, physical values presented in the scientific literature, and considerations of the specific experiment in progress. Within the limits of the logic of an algorithm, these approximations can be modified and managed. Idealizations and approximations are elastic to a certain degree.

Simulation permits the opening of two windows onto temporalities. Simulators are well positioned to explore dynamical properties of phenomena. They are free to introduce intervals of temporality into their numerical experiments that change the reciprocal values of entities. In the domain of thin oxide films, for example, simulation can introduce in a controlled way changes in form, scale, force relations, the position of molecules and atoms, and their state (polarity and energy) across time. One observes that the amount of attention given to questions of dynamics in simulation research is appreciably greater than in metrology-driven experimentation. Difficulties of control in metrology experimentation are far greater.

The second temporality deals with efficiency and speed, that are often equated with simulation. Computers can now carry out big and complicated calculations in relatively short periods. The preparation of algorithms and the introduction of selected quantitative data can also be done relatively quickly. Simulators frequently boast that their endeavors almost always outpace those of metrological experiments.

Simulators experience two types of limitation. Firstly, in spite of everything, computing power is not infinite. Secondly, the logic contained in algorithms defines and restricts what can be expressed and studied in a simulation finding. In thin-film oxide, emphasis is placed on periodicity—this may even be periodicity of defects. To a certain degree, the search for regularity is paramount. This contrasts with metrology experimentation, where practitioners keep an eye out for local strangeness—whatever does not fit into the anticipated pattern.

In experimental investigation of thin oxide films at the nanoscale, the instrument revolution offers new visual and observational possibilities. Metrology can now explore phenomena atom by atom and molecule by molecule, thanks to different kinds of scanning probe microscopy (STM and AFM, see Chapter 1). They can determine the presence or absence of a particular single atom, whether one atom is located on top of another atom, next to it, or comprises a bridge between two other atoms, etc. Based on relative position, metrology experimenters can deduce the forces at play. Beyond this, scanning probe microscopy gives scientists the sensation that they "see" individual elements in their sample. These elements exhibit an immediacy of presence. They are not perceived as mediated. Now, observation of atoms located on thin films means something very different from the collective diffraction points available in earlier categories of detection associated with diffraction and scattering.

Scientists see the surface and shape of nanoscopic objects, their materiality, and almost their palpability. This immediate and compelling quality of experimental observation differs from the idealized entities of simulation representations, where something of the object disappears. To quote one of the team senior scientists:

> Scanning probe microscopy is another step forward. One can walk the tip across the surface of the research-object and obtain an image atom by atom; one can see a group of atoms—a real lattice. If an atom is missing, we see a hole, and if an atom is added, it is visible . . .[53]

The observational perspective concentrates on the local. The local perspective generates decisive information about structures—in the case of thin oxide films, surface structures and relations between surface and environment. This stimulates reflection on the behavior of crystal surfaces and the dynamics of their dissolution.

The above claim of simulators that they are often free to work at an accelerated pace compared with metrology experimenters is not without truth. The director of the metrology-experimentation unit of the laboratory engaged in thin oxide film research that we investigated complained that it proved slow to be equipped—requiring several years. Experimenters had to assemble expensive, complicated, diversified instruments to perform their work. They then had to master the devices and to prepare them for their specific experiments. Finally, time was required to design the experiments. The time needed to conduct experiments is also considerable. There thus exists a great temporal lacuna between the relatively short time-scale of simulation and the comparatively long scale of metrology-based experimental work. This is in part due to several sorts of inertia. One consists of the often slow negotiation entailed in obtaining properly tailored research-objects. As indicated above, this includes the process of modifications in the samples. Changing the settings of the different parameters of a particular experimental run (pressure, temperature, magnetic and electric fields, wetness, etc.) is most of the time very delicate and time consuming.

While the cognitive objectives and work practices of simulators and metrology experimenters are often conspicuously different, this nevertheless leaves open the possibility of cooperation and synergy under particular conditions. As observed in our study, there are three permutations of relations between simulators and metrology experimenters: (1) simulation led, (2) metrology led, (3) simulators and metrology experimenters working in tandem.

2.3.1 Simulation-led enquiry

Particularly in new domains of research, such as are observed in nanoscale surface physics and notably in the case of the exploration of oxide, simulation frequently points in

[53] Interview of Jacques Jupille (INSP, University of Paris) by authors, 22 October 2007.

a direction that subsequently orients experimental studies. This occurs in a framework of prediction and it ultimately offers an explanation of metrological instrument experimental findings. At this scale, there exists a paucity of information, which makes the design and implementation of metrology experiments quite problematic. Simulation has allowed rapid exploration of the forces and entities characteristic of nanosurface physics oxide materials. In simulation, atoms can be shifted to the right or moved to the left and can be arranged in a variety of patterns. Alternative forces can be introduced, and their intensities increased or decreased. Simulation allows numeric experimentation that is unattainable by metrological experimentation. For example, the Schrödinger equation can be solved for very high energy levels; this solution is out of the reach of metrology-based experimenters.[54] This offers the possibility of understanding a precise question (in this case the energy level of a particular surface of ultra-thin oxide) in a broader framework.

Through this play with hypothetical situations, simulators can determine what is physically coherent or incoherent. It establishes what is plausible. Based on plausibility, simulation predicts the outcomes of particular atomic and molecular configurations and forces. For example, in the recent past, it has not been possible to construct an experiment in which a gold atom could be deposited on an oxide surface. Although this experiment is now possible, it remains highly complicated. Conversely, the simulation *ab initio* method has proven highly effective in identifying the position of equilibrium of the gold atom on the surface. This anticipatory knowledge has made it easier for experimenters to design and conduct research projects. This episode illustrates a posture in which simulation opens the way for metrological experimental initiatives to occur at some undefined time in the future. Another routine aspect of simulation work is pure prediction. In the 1990s, the question of polarity was quite well understood for surfaces of bulk materials. Now, with the advent of nanostructured thin-film materials, a range of supplementary questions has arisen. Low-dimensional materials have consequences on surface phenomena that are different from those on the surfaces of bulk substances. Simulation techniques deal with these questions and have generated an ensemble of predictions, which call for experimental validation. Research on stability and electronic structure of polar surfaces at the nanoscale using simulation predicts that they are affected by the chemical material environment, which includes different expressions of titanium di-oxide and strontium mono-oxide. This prediction required the examination of parameters that are not easily accessible to experimenters and that are difficult to explore, such as Fermi bands.[55]

The centrality of prediction in simulation is illustrated by the following quotation:

> In nanoscience, in our domain, simulation precedes experimentation because it predicts results that will hopefully be validated by metrological experimentation. It may be a slight exaggeration to put it this way, but one can say that the experiments (metrology) are

[54] Interview of Jacek Goniakowski (INSP, University of Paris) by authors, 3 May 2008.

[55] F. Bottin, F. Finocchi, C. Noguera (2005) Facetting and (nx1) reconstructions of $SrTiO_3(110)$ surfaces, *Surface Science*, 574(1): 65–76.

constructed around simulation predictions. Simulation predictions establish the axis, the impulse, the experimental orientation.[56]

The fact that prediction is central does not weaken our claim concerning the place of descriptivism in NSR. There is no strong reason for the one to deny the other. The co-existence of prediction and descriptivism are particularly palpable in their encounter. They are certainly not antithetic. It is not an either/or situation.

Another orientation of simulation-led research focuses on questions of morphology, spatial relations, and force. More precisely, simulation-based nanoscale research on oxides deals with five fundamental parameters: (1) the relative positions of atoms and molecules, (2) their shape, (3) environment, (4) energy, and (5) dynamics. The operation of these five components in a simulation research project is reflected in the following example.

Research is carried out on a model crystalline lattice and the effects of selectively cutting it, with particular attention to the position and the re-positioning of individual atoms. Figure 2.3 shows how the initial shape of a crystal determines the geometric transformations in the process of its growth.

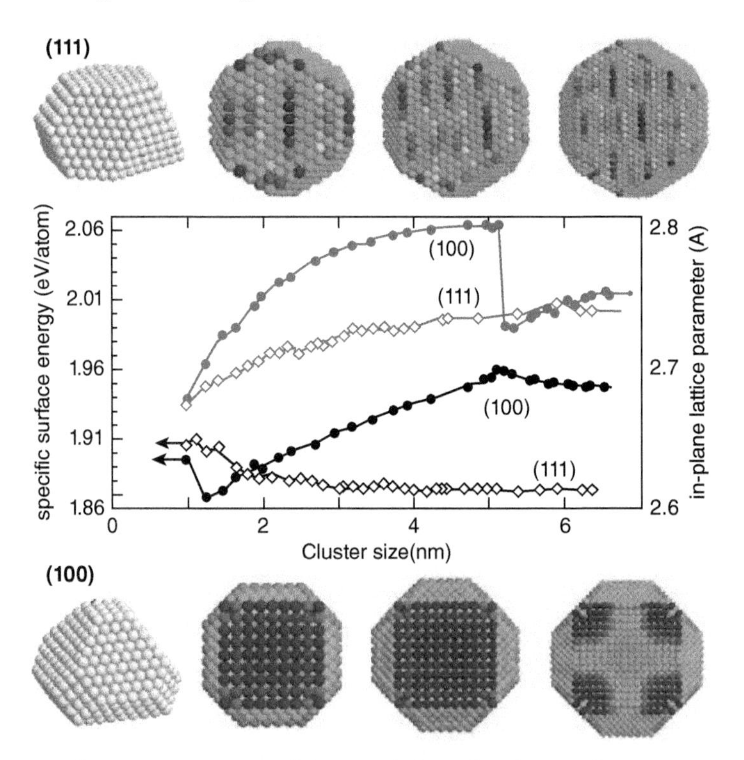

Figure 2.3 Simulation representation of the morphological evolution of two MgO nano-objects whose initial forms were dissimilar, the first (111) and the second (100). Image reproduced from J. Olander, R. Lazzari, J. Jupille, B. Mangili, J. Goniakowski, and G. Renaud, *Phys. Rev. B* 76 (2007) 075409, with permission of the INSP.

[56] Interview of Jacek Goniakowski (INSP, University of Paris) by authors, 3 May 2008.

> Let's imagine that one displaces atoms on a crystalline lattice. You cut this lattice. Beforehand, atoms were surrounded by many neighbors. After the cut, there are atoms that lose many neighbors and others that lose almost none. What are the most reactive atoms? Some individuals are isolated, and some less isolated. Those that are the most isolated are the most reactive. Surface sites react most strongly with the environment and are most likely to receive impurities. If you take away an atom's neighbor, it is discontent because it wants a surrounding environment as before. An atom on a corner, an atom on an edge, and an atom on a surface do not have the same number of neighbors. From the perspective of their characteristics, they are very different . . . We have different kinds of information, position, distance, angles, etc., and in addition to that, information on energy. For example, how much it costs to replace atom A with atom B according to the site. In the corner site, it costs X, in the edge site, it costs Y, etc. We obtain tendencies. It costs less energy to replace an atom on such and such a site, atoms will perhaps all fill that site first. When those sites are fully occupied, the atoms occupy other sites, etc. We thus get a view of how the object is constructed. Once we have this view, we can then calculate optical, excitation, or infra-red properties . . .[57]

This physicist closed the interview with the remark that this kind of calculation is specific to nanoscale objects, contrary to bulk objects, where every element is "identical."

From the above, we see that shape, size, position, force, environment, etc. constitute the landscape of simulation. Numerical calculations of these parameters are bounded by the simulator's background knowledge of the object. This physicist operates in a logic of constraints. One strength of simulation lies in the freedom to introduce values and configurations that stretch the envelope of conventional organizations and representations of objects. In so doing, new constraints may arise, as well as credible unanticipated systems. In this way, simulation switches back and forth between a terrain of logic and a kind of playfulness.

2.3.2 Metrology-led investigation

The development of instrumentation for experimental investigation at the nanoscale permits new questions or allows the framing of older questions in new terms. The possibility of observing new objects at this scale determines topics and the landscape and language of observation. In the physics of ultra-thin surfaces, metrology-equipped experimenters are now capable of identifying the position of individual atoms and the organization of clusters of atoms, and they can describe features of single molecules. When these scientists communicate their observations, they often speak in terms of the size, shape, relative spaces, and textures of materials. In this research work, however, characterization of entities is not the product of a calculation but is instead the result of observation and measurement. As will now be shown, experimenters' vocabulary of description differs sharply from that of simulators.

[57] Interview of Fabio Finocchi (INSP, University of Paris) by authors, 16 April 2008.

When a surface of oxides observed at the nanoscale is portrayed by scanning probe microscopy using for example an AFM, it exhibits a totally chaotic landscape of mountains, plains, and valleys, marked by sharp edges, terraces, or steep steps. Such a description reminds one of the reports of Galileo when first gazing on the surface of the moon. This is the characteristic language used by metrology experimenters to describe and to understand surfaces at the nanoscale. Different aspects of forms and position are employed by these scientists to deduce the presence or absence of particular forces. In experimentation, forces are inferred from form and position. This can provide some indirect information on dynamics.

In surface physics, these morphologies and spaces are registered and captured as images. The images inform the objects—they are the informational foundation and representation of the object; and it is these images that are constitutive of discussion, intelligibility, and communication. The place of images will be discussed at length in Chapter 4. For the time being, we exhibit an image generated in the course of nanoscale research on oxide surfaces.

What scientists observe is a landscape of points, lines, curves, and spaces, which are often complicated and encumbered. Nanoscale surface scientists must next relate items one to another, and this requires selection. One can relate one point to many others, one complex configuration to another; the spacing between atoms in the image or the notion of directionality they suggest induces the researcher to select, to privilege certain links.

Depending on the status accorded to the link, it may be judged to constitute a physical relation. Such relations implicitly refer to some necessary interaction—chemical, electronic, magnetic, etc.

The following quotation demonstrates the centrality of form to observation as revealed through instrument-based images. The research carried out here deals with the dynamics of the dissolution of an oxide in an aqueous environment. In reading this passage, particular attention should be given to the language of description, which reveals key aspects of seeing and thinking of metrology informed experimentation work in this field.

> The MgO (an oxide) sample is placed in an aqueous milieu. One sees that the edges begin to be eaten away by water, but not the corners. At the outset there were cubes. The edges of these cubes are progressively eroded and only then the corners. It is the truncation of the corners that gives rise to tetrahedrons. This truncation, which is what one sees at first. But examine it more closely, it looks rather more like a factory roof. It is not at all round, the truncation corresponds to crystallographic coordinates. One sees a line that is not round, but instead crenelated. It is in the crenulations that lies the question of mechanism.[58]

In this example, mechanism may basically be understood as the chain of objects (forms) and events (transformations) that constitutes the dynamics in a process that goes from A to Z. Here, mechanism lies in the domain of description, and it may perhaps be likened to a film where physical laws play a certain role. This is a clear example of how morphology

[58] Interview of Jacques Jupille (INSP, University of Paris) by authors, 17 September 2007.

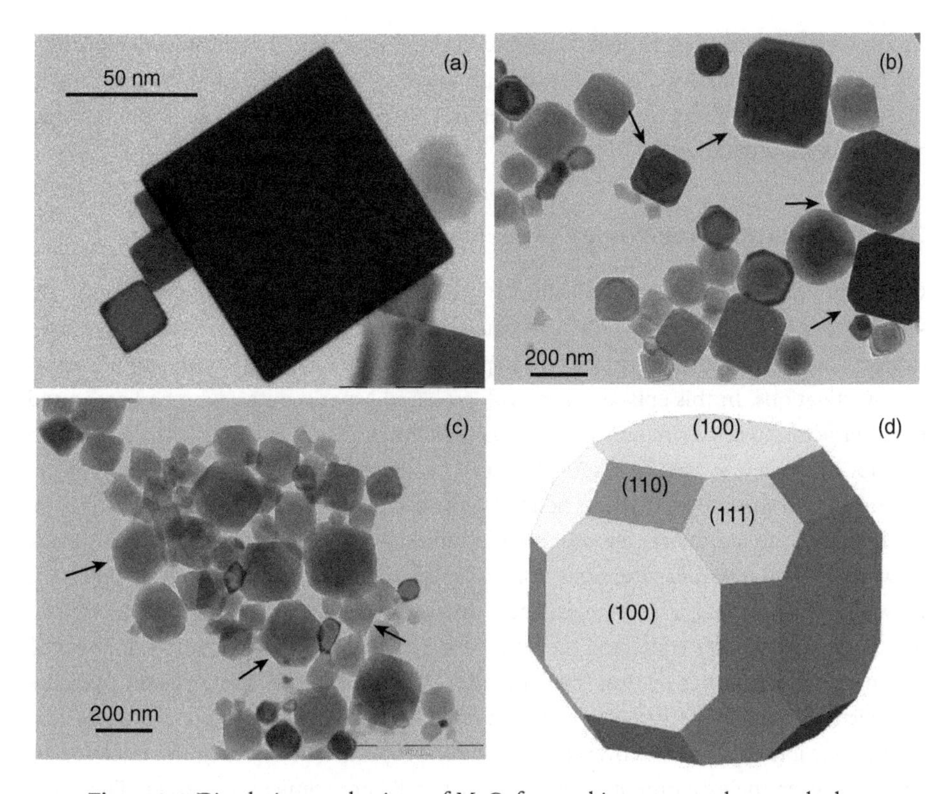

Figure 2.4 Dissolution mechanisms of MgO: from cubic nanocrystals to octahedra.

penetrates the observation of metrology experimenters working in this domain and how it affects language and logic. It is safe to say that the very substance of the questions asked by researchers flows from geometry-driven observation and reasoning.

Figure 2.4 depicts the progressive transformation of an MgO crystal in a process of dissolution. Photograph (a) shows well-defined square shapes exhibiting sharp edges and corners. Photograph (b) reveals changes in the form of the corners which may be likened to shallow stair steps. In photograph (c) the reader can observe a multitude of crystals with different geometries, corresponding to different states of dissolution: one can see almost perfect squares, octahedrons, lozenges resulting from the dissolution of octahedrons, etc.

Meaning is ground in the shapes of objects, and the shapes of objects are inextricably linked to description as understanding. Meaning is progressively developed from the resources of relations. Meaning here signifies the identification of the preferred organization and interactions of selected entities in the system under study. It is strongly related to the special coordinates of a system. This species of meaning is distinct from questions about how a system operates. For these experimenters, meaning refers to what and not how. "How" questions are associated with interpretation. Metrology-instrument experimenters sometimes engage in interpretation of the local system that they have observed.

This is referred to as "experimental interpretation." Beyond issues of experimental interpretation, the plausibility of interpretation frequently belongs, however, to the domain of simulation, as will be seen below, where metrology-instrument experimenters and simulators work in tandem.

2.3.3 Simulation/metrology tandems

Our earlier discussions focused on simulation-led and then on metrology-led research; we now turn to tandem projects. The research incident that we will term "the gray story" explores the contribution of simulation to processes of explanation in nanosurface physics investigations. In this episode, a metrology-based experimenter, who had measured the interface between a zinc/selenium semiconductor and iron, noted that the separation was not well differentiated. In his AFM images, he detected an ill-defined fuzzy region between the two substances that was neither white nor black, as should have been the case. At the nanoscale, atoms can be observed to interlace. Instead of having clear-cut black or white regions, the measurements done on the sample suggested that, at the interface between the two materials, there existed a gray zone consisting of a kind of intertwining. The experimenter contacted the simulator in search of an explanation. In view of the magnetic properties of iron, the simulator decided to attack the problem from the perspective of spin. The structures of iron and the semiconductor matched, and calculating their interface, he discovered remarkable properties concerning the injection of spin. However, the experimenter's measurements did not confirm this result. The simulator next made new calculations and discovered the existence of an interface layer where iron mixed with the semiconductor. The existence of this interface appreciably degraded surface magneto-transport effects. The observed effect was hence not a simple matter of spin, as was initially suggested by simulation. In the "gray story," after the questions raised by metrology and experimental rejection of the opening theoretical simulation explanation, the simulator ultimately provided a subtle explanation of the observed experimental findings, and also contributed to a finer understanding of magneto-transport dynamics.[59]

A last aspect of simulation/metrology work shifts attention back to the opening of this chapter, where it was pointed out that the synthesis of materials in the form of research-objects is crucial to nanoscale physics investigations. Section 2.2 accorded considerable importance, sometimes even center stage, to the technology of synthesis and its dynamic interaction with experiments. It has been demonstrated that linkage between

[59] Interview of Fabio Finocchi (INSP, University of Paris) by authors, 16 April 2008; interview of Rémi Lazarri (INSP, University of Paris) by authors, 24 September 2008. See M. Marangolo, F. Finocchi (2008) Fe-induced spin-polarized electronic states in a realistic semiconductor tunnel barrier, *Physical Review B*, 77(11); F. Finocchi, R. Hacquart, C. Naud, J. Jupille (2008) Hydroxyl-defect complexes on hydrated MgO smokes, *Journal of Physical Chemistry C*, 112(34): 13226–13231; P. Geysermans, F. Finocchi, J. Goniakowski, P. Hacquart, J. Jupille (2009) Combination of (100), (110) and (111) facets in MgO crystals shapes from dry to wet environment, *Physical Chemistry—Chemical Physics*, 11(13): 2228–2233.

experimenters and epitaxy often sophisticates and accelerates the synthesis of materials. The same can be said for interaction between simulation and development of epitaxy methods and performance. Epitaxy is certainly far from a low-theory field. Theories of the best conditions governing crystal growth and the utility and inconvenience of defects have considerably profited the science of nanoscale synthesis. Crystal growth entails insights into the energetics of structural composition and change, and this domain is precisely one kingdom of simulation in nanoscale surface physics. For example, the kinetic Monte Carlo method has been used to calculate growth mechanisms of a material like MgO with reference to surface roughness, size distribution, density of the islands, and filling ratios of the growing layers. It was calculated that the best growth occurs in an environment at above 700 K and a pressure of 0.1 Torr.[60] Such simulation findings often prove important for the practice of epitaxy.

CONCLUSION

Our study of nanoscale physics research prompts three observations. First, in physics, research is conducted with reference to the size, shape, position, texture, and interaction of atoms and molecules. These features constitute the privileged language of observation and are central to intelligibility. Description is of foremost importance and argumentation and proof are often structured around it.

Second, synthesis of materials constitutes the central axis of nanoscale physics, onto which are grafted instruments, methods, questions, concepts, and skills, as well as the above-mentioned landscape of size, morphology, and position. Synthesis initially entails the fabrication of generic nanostructured materials and then it entails the labor of materials-by-design, where a research-object is built in response to the particular needs of a specific experimentation project. This draws attention to the capacity for control, which is a constant in nanophysics. Through precise tailoring of a substance, physical properties, previously unobserved, are produced and explored. Research-objects function as a bridge between synthesis and metrology experimentation, and in so doing they comprise a key combinatorial of NSR.

Third, in nanophysics, one can identify three principal cognitive, instrumental and technical groups: metrology-based experimenters, epitaxiors, and simulators. Each of these populations possesses specific skills. One characteristic of nanophysics may reside in the elevated amount of communication, even interlacing between the three bodies. We hypothesize that, due to the combinatorials of metrology / simulation collaborations and epitaxy / experimentation combinatorials (the hole underpinned by synthesis), synergy is possibly greater in nanoscale research than in other fields.

[60] E. Antoshchenkova, M. Hayoun, F. Finocchi, G. Geneste (2012) Kinetic Monte-Carlo simulation of the homoepitaxial growth of MgO{001} thin films by molecular deposition, *Surface Science*, 606(5–6): 605–614.

3

The Scale of Life?

Research in nano-related biofields has grown considerably in the course of the last twenty years. Characteristics of bio-nano studies emphasize investigation at the level of a single molecule. Nanoscale biology contributes to the understanding of larger objects through examining them in terms of molecular components and behavior. Molecular control is of foremost importance here. The notion of control comprises a transverse feature. In this chapter we address two issues. What research questions and practices underpin the laboratory efforts of scientists who identify themselves with nanobiology? What new specific areas of knowledge are emerging in these fields?

The documentation and explanation of the exponential growth of research in nanobiology since 1990 is the topic of the first section, entitled "A new effervescence." The second section, "The reach of instrumentation," explores how some of the instruments presented in Chapter 1 operate in a specific way in nanobiology. Nanobiology instrumentation proves particularly remarkable for its range and intensity of combinatorials. Numerical simulation is crucial both to experimentation and to the organization and exploitation of biodata banks that are so central to nanobiology. The third section of this chapter, "Molecular biology nano style," presents the six components crucial to bio-nanoscale research: structure/form, binding, function, three-dimensionality, environment, and control. The specificity of nanobio resides in the particular relations that exist between these elements.

"Nanobiology research at work: from codes to configurations," the fourth part, investigates several techniques for the production of pre-designed bio molecules. In the logic of molecular design, the production of new artificial nanobio materials is a central orientation. Reflection on the significance of the specificity of the form of protein molecules constitutes the theme of the fifth section, entitled "Life as a dynamic Lego game." Protein segments are frequently referred to as Lego by scientists. The Lego perspective introduces dynamics into the comprehension of biological processes at the nanoscale, and particularly with reference to mechanisms of the evolution of life. The sixth section, "Binding, sensing, and detecting," shows that in the biophenomena of binding, sensing, and detecting, combinatorials are visible in the conjoining of function and structure.

3.1 A NEW EFFERVESCENCE

A remarkable discontinuity in the quantity of biology publications on molecular objects occurred in the early 1990s.[1] In the two decades between 1990 and 2010, the annual number of published articles relating to many biological substances grew by between roughly twenty- and fiftyfold: in the case of protein and folding topics, annual publication has climbed from 90 in 1990 to over 2000 in 2012; for protein and function, from 1200 in 1990 to over 47,000 in 2012; for DNA and design, from 80 in 1990 to over 5200 in 2012 (see Figures 3.1 to 3.3).

To imply that this expansion is uniquely due to research activity in nanobiology would be misleading, although it certainly does constitute a principal component. The ascending trajectory of nanobiology research is connected with the conjunction of two distinct elements. A first crucial factor is institutional / financial. The International Genome Project was launched in 1988, principally by the American National Institute of Health and the Department of Energy. Funding of biology-related research rocketed in the USA. Spending stood at \$27.9 million in 1988, at \$86.7 million in 1990, rising to \$134.8 million in 1991, and to \$437 million in 2003, the year when the human genome code was totally identified. Finance in Canada, Japan, the UK, France, Sweden, China, Korea, and Australia followed suit. In 2000, global spending attained \$1,805,325,000.[2] The launching

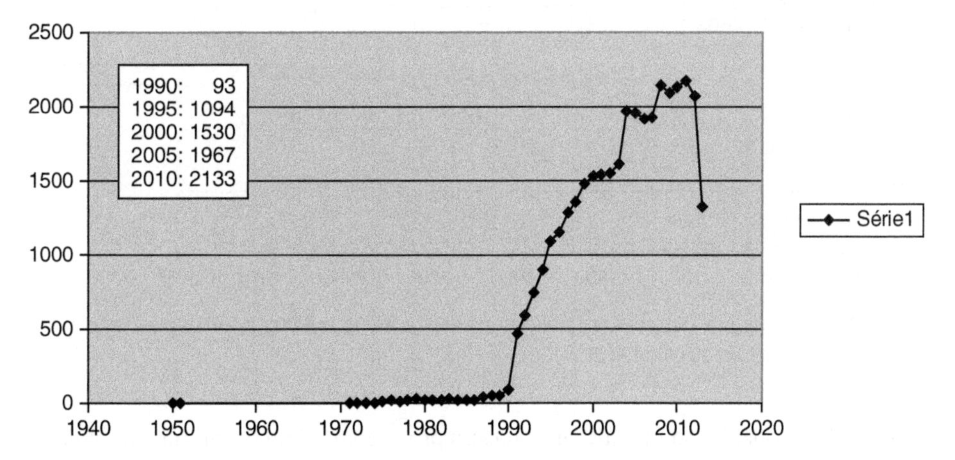

Figure 3.1 Protein folding. Number of publications: topic: protein* and folding*: 35,400 items (Thomson ISIWeb of Knowledge: 10 October 2013).

[1] The term "nanobiology" is current in biology-related and medical-science-related publications. The growth of this field has received considerable attention by the scientometrician sociologists of science. See for example: A.L. Porter, J. Youtie (2009) How interdisciplinary is nanotechnology? *Journal of Nanoparticle Research*, 11(5): 1023–1041; H. Eto (2003) Interdisciplinary information input and output of nanotechnology project, *Scientometrics*, 58(1): 5–33; J. Schummer (2004) Multidisciplinarity, interdisciplinarity, and patterns of research collaboration in nanoscience and nanotechnology, *Scientometrics*, 59: 425–465.

[2] <http://www.stanford.edu/class/siw198q/websites/genomics/entry.htm> (consulted 22 March 2011).

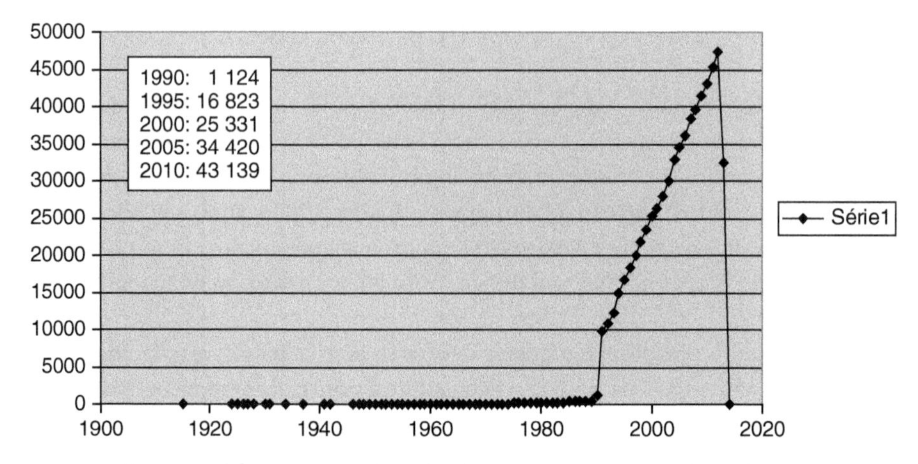

Figure 3.2 Protein and function. Number of publications: topic: protein* and function*: 654,522 items. (Thomson ISIWeb of Science: 10 October 2010).

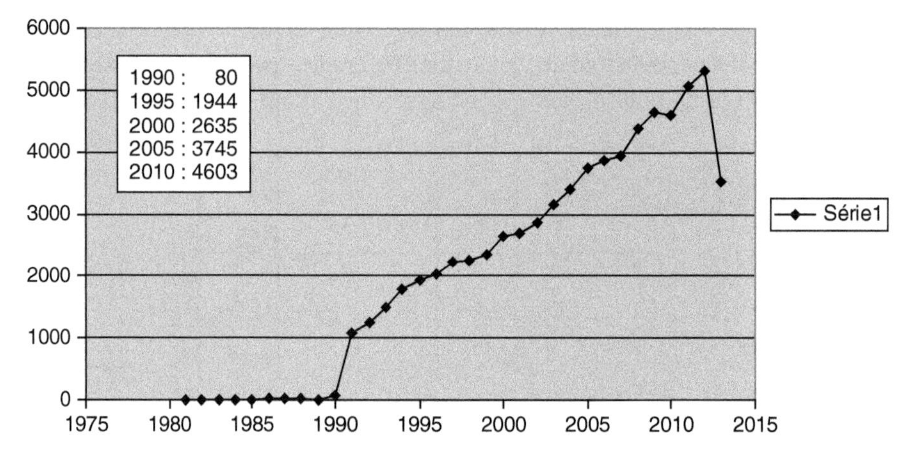

Figure 3.3 DNA design. Number of publications: topic: DNA* and design*: 70,408 items (Thomson ISIWeb of Science: 10 October 2013).

of the US National Nano Initiative in 2000 also provided funding specifically for biology. The overall budget in 2009 stood at $1.5 billion. Again in the United States, the National Science Foundation similarly earmarked money for nanobiology exploration.

As indicated by some of our interviewees, some physicists came to see in biological materials a new turf for extension of their research, and this constituted a second important factor in the growth of biological investigations in the nano style during this decade.[3]

[3] O. Bueno (2011) When physics and biology meet: The nanoscale case, *Studies in History and Philosophy of Science*, Part C, 42(2): 180–189; A. Marcovich, T. Shinn (2011b) The cognitive, instrumental and institutional origins of nanoscale research: The place of biology. In: M. Carrier & A. Nordmann (eds.), *Science in the Context of Application*. Springer Netherlands, pp. 221–242.

The research trajectory of Philippe Minard illustrates this encounter. Minard is a biochemist (Institut de Biochimie et Biophysique Moléculaire et Cellulaire, University of Orsay), who currently works in the field of protein engineering. He recounts that during the late 1990s, the dominant perception among molecular biologists was that nothing more significant was to be learned about the functioning of life at the molecular level through images, such as electronic microscopy, NMR, or crystallography. It was only when physicists requested delivery of biological molecules for their own instrumental investigation with an STM and AFM, that awareness of the possibilities for exploring more deeply the functioning of complex biological molecules arose. In the case of Minard, this occurred when physicist colleagues at the Ecole Normale de Lyon, solicited biological materials from him for their own physics based experiments. The images that these physicists produced from the proteins that Minard had provided, opened new horizons of research for him. In effect, this episode redirected Minard's research trajectory toward nano style projects; he now conducts investigations on the morphology of enzymes and their function, using AFM, optical tweezers, and numerical simulation.[4] The growth in nano publications related to bio materials can be explained by initiatives among material scientists: as early as the 1990s, biological substances such as peptides (a short strand of a protein), were being used as templates for fabricating, for example carbon nanotubes, because of their self-assembling characteristics and for the architecture they could offer.[5]

The third important factor is the growth of nanobiology research. During the mid to late 1980s, many families of new metrological instruments became available that directly or indirectly suited biological research (see Section 3.2). The extension of bio-nano took advantage of the multiplication of combinatorials that particularly benefited biological investigation on the nanoscale. This is notably true in the case of optical tweezers.[6] Some nano instruments were specifically conceived, designed, and commercialized for nanobiological work. In parallel, computational research developed both in terms of hardware and software. One important confluence between metrological instrumentation and computational instrumentation resides in the capacity of both to represent molecules in three dimensions, as virtual entities, and as materials.

3.2 THE REACH OF INSTRUMENTATION

The dramatic post-1990 acceleration in nanobiology research is the product of synergies between different fields of investigation and combinatorials between new instruments

[4] Interview with Philippe at Institut de Biochimie et Biophysique Moléculaire et Cellulaire, University of Orsay (IBBMC), by authors, 12 June 2010. Another example of this encounter between physicists and biologists is described in our discussion of the work of James Gimzewski (see Chapter 6).

[5] M.R. Ghadiri, J.R. Granja, R.A. Milligan, D.E. Mcree, N. Khaznovich (1993) Self-Assembling Organic Nanotubes Based on a Cyclic Peptide Architecture. *Nature*, 366 (6453): 324–327.

[6] Optical tweezers (originally called "single-beam gradient force trap") are scientific instruments that use a highly focused laser beam to physically hold and move microscopic dielectric objects. Optical tweezers are employed successfully to study biological systems. <http://en.wikipedia.org/wiki/Optical_tweezers> (consulted 30 April 2011).

and methods, and new ways of seeing bio molecules. In the pages that follow, we examine the profusion of novel instruments that have contributed to the growth of metrology and simulation-driven experimentation in nanoscale biofields.

Before addressing the issue of instrumentation, however, it is essential to describe the foundational methodology that constitutes a necessary precondition for much of nanobiology. The molecular biology method, polymerase chain reaction (PCR)[7] allows the mass production of pre-selected or designed nucleic acids (both DNA and RNA) sequences. PCR lets scientists create desired molecules of DNA of specified length and form, or possessing the necessary base ordering for generation of the intended protein sequence. It is on these indispensable objects that nanobiologists develop their research. This method, which was first introduced in 1990 and fully described in 1993, permits the amplification or copying (of order about one billion) of an already known DNA or RNA sequence from a very small quantity of nucleic acids. One of the most interesting applications of the PCR method, directed evolution, is a method used in protein engineering to harness the power of natural selection to evolve proteins or RNA with desirable properties not found in nature (see Section 3.5).

3.2.1 Metrological instrumentation

The dramatic rise in publication of nanoscale-related research after 1990 is attributable to the instrumentation revolution of the 1980s and to the combinatorials resulting from integration of new instruments with previous generations of devices. Chapter 1 described the invention and centrality of the STM (1981) and the subsequently invented AFM (1985). The AFM is a leading metrological instrument present in much nanoscale biological research. As indicated in Chapter 1, in contrast with the STM, the AFM can measure non-conducting materials. An AFM provides the ability to perform three-dimensional measurements of surface structures at a few nanometers' scale in an ambient and liquid environment. The capacity to measure objects at this scale in an aqueous milieu is highly important, because it corresponds with the vast majority of biological conditions. This is an advantage over other instruments such as the electron microscope, which requires often destructive sample preparation, most notably by drying and placing them in a vacuum.

The AFM offers important possibilities for the investigation of DNA condensates. It is capable of viewing the structure of the delivery vehicle in its hydrated state, as it occurs in cells. On the basis of AFM nanoscale resolution, one can image DNA strands and see how they react and condense with a particular polymer. Among other applications, the AFM is a gene-delivery vehicle.[8] It can observe in real time nanoscale processes inside in vivo materials. This possibility is recent in biological research and particularly pronounced

[7] P. Rabinow (1996) *Making PCR. A story of biotechnology*. Chicago, IL: University of Chicago Press.

[8] L. Wilson, P.T. Matsudaira, B. Ramaswamy, J.H. Horber (2002). *Atomic force microscopy in cell biology* (Vol. 68). Access Online via Elsevier.

in nanoscale-apparatus measurement capability and research findings. It will be shown below that this potential is reinforced through combinatorials in which additional devices like spectroscopy and lasers operate in tandem with quantum dots.

Three additional instruments that developed *c.*1990 are also particularly current in nanobiology work. Near-field optical microscopy, capable of nanometric measurements of bio materials in their natural environment, arose during the same period.[9] The limit of optical resolution in a conventional microscope, the so-called diffraction limit, is on the order of half the wavelength of the light used to image. Thus, when imaging at visible wavelengths, the smallest resolvable objects are several hundred nanometers in size. Using near-field optical techniques, researchers currently resolve features on the order of tens of nanometers. This makes it possible to study large macromolecules and assemblies of molecules.

A brief mention of two further recent and particularly nanobio-relevant instruments is warranted before moving on to a discussion of combinatorials between newer and older devices. A new form of mass spectroscopy constitutes a great improvement over earlier devices which were far less sensitive. The novel nano-sensitive spectroscopy system was invented at the Japanese instrumentation firm, Shimadzu Corporation, by Koichi Tanaka, 2002 Nobel Prize in chemistry. This spectroscopy system was specifically conceived for biology research and was viewed as the wave of the future. Tanaka's project, metric-assisted laser desorption ion time of light spectroscopy (MALDITOF),[10] began in 1987 and was commercialized in 1993. It constitutes a cornerstone of today's nano bio instrumentation revolution.[11] The second is the controlled production of quantum dots (see Chapter 1), which are now used as biological markers. These were developed at Texas Instruments in 1987, and the devices started to be manufactured in commercial quantities in 1993.[12] Quantum dots are today a device that is widely employed for exploration of the internal structure, and particularly the internal dynamics, of bio materials (see Chapter 5). They can be selectively attached to specified components inside molecules and they possess physical properties that allow them to absorb and emit light at different, predictable wavelengths, according to their size. This permits nanobiologists to study intra-molecular dynamics. Quantum dots are widely used in different areas of nanobiology imaging because they can track molecular movements.[13] The AFM, near-field optical

[9] H. Novotny (2007) The history of near-field optics. In: E. Wolf (ed.) *Progress in Optics*. Amsterdam: Elsevier, pp. 137–184.

[10] Matrix-assisted laser desorption/ionization (MALDI) is a soft ionization technique used in mass spectroscopy, allowing the analysis of biomolecules and large organic molecules, which tend to be fragile and fragment when ionized by more conventional ionization methods. <http://en.wikipedia.org/wiki/Matrix-assisted_laser_desorption/ionization> (consulted 30 May 2013).

[11] K. Tanaka (2003) The origin of macromolecule ionization by laser irradiation, Nobel lecture, 8 December 2002. *The Nobel Prizes 2002*, Stockholm: Tore Frängsmyr.

[12] M. Reed (1993) Quantum dots. Nanotechnologists can now confine electrons to point like structures. Such "designer atoms" may lead to new electronic and optical devices, *Scientific American*, 268(1): 118–123.

[13] Interview of Bertrand Tavitian (CEA "Imagerie de l'expression des gènes," INSERM U 803, France) by Anne Marcovich 13 November 2007. <http://www.savoirs.essonne.fr/dossiers/la-vie/medecine-sante/imagerie-moleculaire-une-revolution-en-cours/complement/resources/> (consulted 12 December 2012).

microscopy, MALDITOF, and quantum dots have proved particularly amenable to nanobiological investigations because, as said previously, they do not require damaging sample preparation. For nanoscale biological research, the power of all of these instruments lies in their capacity to study biological materials in ambient conditions: room temperature, pressure, humidity, chemical environment(such as neutral pH) etc. In addition, older techniques such as X-ray diffraction crystallography,[14] necessitated hundreds of thousands of individual measurements to generate a single image, and this took many years to complete.[15]

Three initially biology-unrelated devices have, during the last two decades, intermeshed with new-generation instruments to form novel and powerful combinatorials. The older devices include laser technology and fluorescence-marking techniques. The laser was developed during the 1950s and 1960s. It was initially a theoretical device that was materialized in 1952 (Alfred Kastler (1902–84), Nobel Prize in physics 1966). The concept of optical pumping[16] lies at the heart of contemporary laser technology.[17] The laser can be considered a generic device. Nowadays, it is used for the purposes of measurement and of technological control, and operates as a leading research tool in nanobiotechnology. It frequently functions in combination with quantum dots. This combinatorial of quantum dots and lasers proves essential to studies of location and motion on DNA strands, where the scale is so tiny that the slightest optical aberration will invalidate the results (see Chapter 5).

Fluorescence is relevant to biology in the framework of what is termed fluorescence resonance energy transfer (FRET).[18] Fluorescence is a physical phenomenon, today associated with lighting. It is used as a marker in studying relative positions and dynamical processes in the sciences and particularly in biology, where its potential was first established in the 1930s. FRET is a mechanism describing energy transfer between two parts of a molecule. A donor segment, initially in its excited electronic state, may transfer energy to an acceptor segment through nonradiative dipole–dipole coupling. This circuit is

[14] X-ray diffraction devices, born in 1910, were not applied to biology until the 1930s because the technique required crystallization of the observed object. Only in 1937 did Dorothy Hodgkins introduce techniques for the crystallization of proteins. Nevertheless, observation of proteins proved so difficult and slow that the techniques did not bear fruit for some twenty years. In the 1950s and 1960s, the obstacles inherent to X-ray crystallography technology continued to be such that high resolution protein study still remained difficult. This meant that very few substances were analyzed (E. Francoeur, 1997). The forgotten tool: The design and use of molecular models. *Social Studies of Science*, *27*(1), 7–40). Only three proteins were described (myoglobin and hemoglobin, in 1955, and later insulin, in 1968). These results were considered so ground-breaking that John Kendrew and Max Perutz shared the 1962 Nobel Prize in chemistry for their work on myoglobin and hemoglobin (Cl. Debru (1983). *L'Esprit des protéine*).

[15] Cambrosio, A., D. Jacobi, P. Keating (2005) Arguing with images, Pauling's theory of antibody formation, *Representations* (Journal of Digital Publishing), 89(1): 94–130.

[16] Optical pumping is a process in which light is used to raise (or "pump") electrons from a lower energy level in an atom or molecule to a higher one. It is commonly used in lasers, to pump the active laser medium so as to achieve population inversion. The technique was developed by 1966 Nobel Prize winner Alfred Kastler in the early 1950s. <http://en.wikipedia.org/wiki/Optical_pumping> (consulted 24 April 2007).

[17] J.L. Bromberg (1991) *The Laser in America. 1950–1970*. Cambridge, MA: The MIT Press.

[18] <http://en.wikipedia.org/wiki/F%C3%B6rster_resonance_energy_transfer> (consulted 4 March 2011).

efficient in the determination of small distances and donor/acceptor relative movements. The FRET instrument is often used in combination with an AFM nanoresearch device.[19]

We will close this metrology instrumentation section, which documents the linkage between nanoresearch apparatus and older families of devices, with two emblematic nanobiology research projects. In a first example, the project consisted of detecting the TrkB enzyme in a neurological tissue.[20] In this research, a quantum dot was attached to a specific enzyme, where it functioned to identify the enzyme's location in nerve tissue. Here biomarkers were imaged by the combination of an AFM and a confocal laser scanning microscope (CLSM),[21] providing resolved (nanometer scale) structural and fluorescent imagery. In this example, the quantum dot serves as a marker and the CLSM operates as a detector, while the AFM measures properties of the target substance. To this ensemble must be added fluorescence. When stimulated by a laser beam, quantum-dot light emission takes the form of fluorescence. The sophistication of each technical link in this research process calls on the sophistication of other technical links, and this constitutes a kind of ever higher technical complexity that permits researchers to see more detail and ever finer dynamics of the objects under study.

A second, even more illustrative example of combinatorials between metrological instruments can be seen in a research project that aligns eight instruments: quantum dots, AFM, CLSM, field-emission scanning electron microscopy (FESEM), transmission electron microscopy (TEM), X-ray photo-electron microscopy (XPS), Fourier transform infra-red spectroscopy (FTIR), and optical emission spectroscopy (OES). In the course of this research, it was learned that a substance habitually used to coat quantum dots is more cytotoxic and less fluorescent than an alternative chemical. This research thus extends the chain of sophisticated instrumentation techniques, which in turn allows exploration of complexities at the nanoscale in biology.[22] The conclusion is that developments in entirely novel metrological instrumentation and the emergence of new instrument combinatorials in the rise of nanobiology are inescapable. It is through such devices that single-molecule observation and analysis has become possible during the last twenty

[19] W.J. Greenleaf, M.T. Woodside, S.M. Block (2007) High-resolution, single-molecule measurements of biomolecular motion, *Annual Review of Biophysics and Biomolecular Structure*. Book Series: *Annual Review of Biophysics*, 36: 171–190

[20] J.W. Park, A.Y. Park, S. Lee, N.K. Yu, S.H. Lee, B.K. Kaang (2010) Detection of TrkB Receptors Distributed in Cultured Hippocampal Neurons through Bioconjugation between Highly Luminescent (Quantum Dot-Neutravidin) and (Biotinylated Anti-TrkB Antibody) on Neurons by Combined Atomic Force Microscope and Confocal Laser Scanning Microscope. *Bioconjugate Chemistry*, 21(4), pp. 597–603.

[21] Confocal microscopy, introduced in the 1950s and 1960s, is an optical imaging technique used to increase optical resolution. The principle of confocal imaging was patented in 1957 by Marvin Minsky. The technique aims to overcome some limitations of traditional wide-field fluorescence microscopes. Confocal microscopy permits observation of structures situated several layers beneath an object's surface. In 1978, a confocal microscope was coupled to a laser to scan the three dimensions of a surface. This technique is particularly present in nanobiological research. <http://en.wikipedia.org/wiki/Confocal_laser_scanning_microscopy> (consulted 10 June 2013).

[22] J. Pan, Y. Wang, S.S. Feng (2008) Formulation, characterization, and in vitro evaluation of quantum dots loaded in poly(lactide)-vitamin E TPGS nanoparticles for cellular and molecular imaging, *Biotechnology and Bioengineering*, 101(3): 622–633.

years. It is the recent capacity to see inside molecules and to explore dynamical processes, such as folding, that is constitutive of nano in the life sciences.

3.2.2 Computational instruments—simulation

A second category of instrumentation, namely computational instruments that produce simulations of physical phenomena and thereby descriptions and predictions of properties and dynamics, has operated as an additional motor to the dramatic acceleration in nanobiological research in the 1990s and beyond.[23] Simulation has been central to the study of proteins and DNA. As indicated in Chapter 1, advanced computational modeling in the domains of the physical and life sciences expanded in the 1970s, 1980s, and 1990s with the advent of three principal conditions : (1) The invention of the microprocessor and development of enhanced electronic memory, beginning in the 1970s and 1980s.[24] (2) The creation of *ab initio* methods and density functional theory between the 1960s and 1980s, permitting the prediction of physical properties through mathematically driven deductive influences based on the atomic number of elements appearing in the Mendeleev table. (3) The massive diffusion of low-cost, user-friendly, powerful desktop computers in science-research laboratories often requiring no specialized learning.[25] Simulation of biomolecules allows the calculation and prediction of energy levels (see Section 3.3) and other physical characteristics, which in turn affect the internal organization of molecules, sometimes expressed in their structure and form. This has become a key signature of nanoscale research, especially in biology. Based on some equations and semi-empirical evidence, simulations can predict the features and dynamics of biological molecules.

In the 1980s and 1990s, bio-informatics developed three different roles. Firstly, it evolved in such a way that the most challenging horizon consisted of analysis and representation of nanobio-objects such as nucleotides and amino acid sequences, or protein structures and domains. Secondly, bio-informatics now entails the creation, organization, and consultation of databases, algorithms, and computational and statistical techniques to solve formal and practical problems arising from the management and analysis of vast quantities of biological data. The area of bio-informatics matured with the institutionalization of the Human Genome Program. Along with metrological experiments, bio-informatics computer programs made it possible to stock, manage, and manipulate a huge amount of DNA sequence-related data. Major research efforts in bio-databases include sequence alignment, genome assembly, protein structure, protein prediction, prediction of gene expression, the modeling of evolution, etc.

[23] A. Carusi, B. Rodriguez, K. Burrage (2013) Models systems in computational systems biology, in J.M. Duràn and E. Arnold *Computer Simulations and the Changing Face of Scientific Experimentation*. Cambridge Scholars Publishing: 118–145.

[24] C. Lécuyer, D. Brock (2010) *Makers of the Microchip. A Documentary History of the Fairchild Semiconductor*. Cambridge, MA: The MIT Press.

[25] J. Lenhard (forthcoming), Disciplines, Models, and Computers: The path of computational quantum chemistry Disciplines, models and computers. The path to computational chemistry. *Studies in History and Philosophy of Modern Physics*.

A third highly important biocomputation sphere consists of visual molecular dynamics—often referred to as molecular graphics.[26] Computational graphics allows biologists to express as visual representations the structures and dynamics proposed by molecular simulation. Advanced graphics enables researchers to manipulate their representations (rotate them, change the angle of the axis, obtain an alternative angle of observation, penetrate beneath the surface of a molecule in order to better observe packing, etc.) in innumerable ways. Through computer graphics, they can also modify the form exhibited in the representations to determine the extent to which such modifications are compatible with the molecule's internal organization and forces (see Section 3.3 on energy landscape).

Algorithms are of foremost importance. The number of simulation algorithms multiplied considerably during the 1990s and subsequently, and their capacity for multi-fold parameter analysis and precision expanded in a variety of biological areas. This evolution was built on a virtuous circle, where informatics-organized data are collected in coherent databanks, and on the basis of these databanks, simulations are performed whose results in turn extend the databanks.

This dynamic is exemplified by the basic local alignment search tool (BLAST), created in 1990 on the eve of the vast dilation of bio-nanoscale research.[27] BLAST is a frequently employed algorithm that enables comparison of primary biological sequence information, like amino-acid sequences of different proteins to other amino-acid sequences, or the nucleotides of DNA sequences to other nucleotides.[28] Another possibility offered by BLAST permits the researcher to compare a query sequence with a library or databank of sequences, and thus to identify library sequences that resemble the query sequence above a certain probability threshold.

The program, called Rosetta, exemplifies how simulation works and the kinds of knowledge it can generate in nanobiology. The Rosetta program derives much of its descriptive and analytic power from simulation. It was developed for the prediction and design of protein and RNA sequences. Created in 1995 by David Baker at the University of Washington, it was used in 1999 by only three nanobiologists for single-molecule nanoprotein studies (see Section 4.1).[29] In 2009, Rosetta was employed by over one hundred nanobiology practitioners. The many permutations offered by Rosetta permit consideration of huge quantities of complex parameters in a very systematic and rigorous way. The Rosetta program features a three-dimensional representation of proteins, which highlights issues of spatial occupancy (see Sections 3.4 and 3.5 and Chapter 4). It may be viewed as emblematic of the nanometric perspective in the life sciences. The

[26] W. Humphrey, A. Dalke, K. Schulten (1996) VMD: Visual molecular dynamics, *Journal of Molecular Graphics*, 14(1): 33–41.

[27] S.F. Altschul, W. Gish, W. Miller, E.W. Myers, D.J. Lipman (1990) Basic local alignment search tool, *Journal of Molecular Biology*, 215(3): 403–410.

[28] S.F. Altschul, T.L. Madden, A.A. Schaffer, Z. Zhang, W. Miller, D.J. Lipman (1997) Gapped BLAST and PSI-BLAST: A new generation of protein database search programs, *Nucleic Acids Research*, 25(17): 3389–3402. This article has been cited more than 30,000 times.

[29] Interview of Brian Kuhlman (University of North Carolina, Chapel Hill) by authors, 18 June 2010.

authors of the Rosetta program have developed an interactive online game ("Foldit") for the broader public, where players are invited to identify pathways in the extremely complex labyrinth that constitutes a protein. The possible pathways indeed prove so complex that contemporary simulation does not necessarily address adequately issues of folding and binding. It is interesting to note, as attested by scientists, that players of "Foldit" have sometimes identified folding itineraries of proteins more effectively than simulation algorithms.

At the core of Rosetta are potential functions for computing the energies of interactions within and between biomacromolecules. One of the principal aims is identification of the lowest energy structure for a protein or RNA sequence. Simulation feedback from the prediction and design tests is used continually to improve the potential functions and the Rosetta search algorithms. Note that many of these nanobiology simulation instruments are intended to promote understanding and practices where molecular prediction, design, and engineering are the foremost concern. As argued above, these components embody one of the key aspects and aspirations of nanobiology research.

In conclusion, the instrumentation revolutions of the 1980s and 1990s have produced a wave of devices that today permit investigation of biological structures and configurations that can be observed and understood at the nanoscale. In this dynamics, practitioners have attained ever higher degrees of control over molecular objects.

3.3 MOLECULAR BIOLOGY NANO STYLE

The genesis of molecular biology dates back to the 1940s and even before. It is clear that nanoscale research belongs to the broader molecular biology movement. In spite of the fact that nanobiology explores the same objects as molecular biology, the questions asked, the way they are investigated, the manner in which objects are handled, and the importance of control are all different.

The biological objects most central to contemporary research are by virtue of their very scale a privileged terrain of nanoscale research. DNA measures 2.5 nm in width, and the average length of a pair of bases in DNA is 0.33 nm. Amino acids, which constitute the components of proteins and peptides, are also nanometric. Questions that are associated with the origins, evolution, and functioning of life routinely include objects and forces situated at the nanoscale. The famous chemist, Linus Pauling (1901–94), Nobel Prize in chemistry (1954), who worked on hemoglobin during the 1930s and 1940s, acknowledged the necessity to reflect in terms of an ever decreasing biological scale which extended from 10^{-7} meters down to the angstrom.[30] At this very reduced scale, it is possible to

[30] E.F. Keller (1990). Physics and the emergence of molecular biology: A history of cognitive and political synergy. *Journal of the History of Biology*, 23(3), 389–409; L.E. Kay (1993). *The Molecular Vision of Life: Caltech, the Rockefeller Foundation, and the rise of the new biology*. New York, NY: Oxford University Press; Linus Pauling, Nobel Lecture: Modern Structural Chemistry. Available at: <Nobelprize.org> 14 Feb 2013. <http://www.nobelprize.org/nobel_prizes/chemistry/laureates/1954/pauling-lecture.html>; A. Cambrosio, D. Jacobi,

determine by calculation the angles formed between two atoms in a molecule and thus to predict the configuration of the molecule.

As shown in the preceding chapter on nanoscale research in the physical sciences, the nanoscale is inextricably connected with the problem of shape. In biological research, the structures of DNA and proteins are today often depicted and understood in terms of form. Issues of biological control and manipulation of the shape and function of molecules are central, and have been so particularly since the 1990s. Nanoscale control is emblematically expressed in work on protein design and engineering, in the building of scaffolding with DNA or proteins, and in development of entirely new biological structures. Control and manipulation, as well as enhanced power of description of molecules, are in large part the outgrowth of the development of a range of new metrological and conceptual instruments—notably the AFM, super resolution microscopy, nuclear magnetic resonance (NMR) spectroscopy, and mass spectroscopy, as well as computational hardware, software, and novel concepts such as energy landscape theory.[31] Seeing nanometric objects in terms of three-dimensional structures and relations (sometimes in real time) has been made possible for the first time by such instruments. The stylized representations of the pre-1980s are now by and large surpassed in the world of nanobiology.

Investigations carried out in biology on the nanoscale essentially deal with two large categories of objects: (1) proteins and enzymes, or fragments of proteins (peptides) and their amino acids, on the one hand,[32] and (2) nucleic acids (DNA and RNA) on the other. Items such as cells, membranes, organelles (ribosome, mitochondria, endoplasmic reticulum, golgi bodies, etc.) are studied at the nanoscale, uniquely in terms of their components. In the nanoscale perspective, biological objects are observed, manipulated, and discussed in terms of architecture, form, and function.

We suggest that the presence, convergence, and intertwining of six specific elements characterize nanoscale research in biology: (1) structure / form, (2) three-dimensionality,[33] (3) binding, (4) function, (5) environment, and (6) control. These are shown in our nanobiology hexagon (Figure 3.4).

P. Keating (2005) Arguing with images, Pauling's theory of antibody formation, *Representations* (Journal of Digital Publishing), 89(1): 94–130; B. Alberts, A. Johnson, J. Lewis, M. Raff, K. Roberts, P. Walter (2002) *Molecular Biology of the Cell*. New York, NY: Garland (4th ed.).

[31] According to the ISI Web of science: The first article published on this topic was: A. Fernandez, E.I. Shakhnovich (1990) Activation-energy landscape for metastable RNA folding, *Physical Review A*, 42(6): 3657–3659.

[32] C. Debru (1983) *L'Esprit des protéines*. Paris: Hermann; R.M. Burian (1993) Technique, task definition, and the transition from genetics to molecular genetics: Aspects of the work on protein synthesis in the laboratories of J. Monod and P. Zamecnik, *Journal of the History of Biology*, 26: 387–407; E.F. Keller (1990). Physics and the emergence of molecular biology: A history of cognitive and political synergy.

[33] E. Francoeur (1997) The forgotten tool: The design and use of molecular models, *Social Studies of Science*, 27(1): 7–40; M.S. Morgan, M. Morrison (eds.) (1999) *Models as Mediators: Perspectives on Natural and Social Science*. Cambridge: Cambridge University Press; E. Francoeur, J. Segal (2004) From model kits to interactive computer graphics. In: S. de Chadarevian, N. Hopwood (eds.) (2004) *The Third Dimension of Science*. Stanford, CA: Stanford University Press: 402–429.

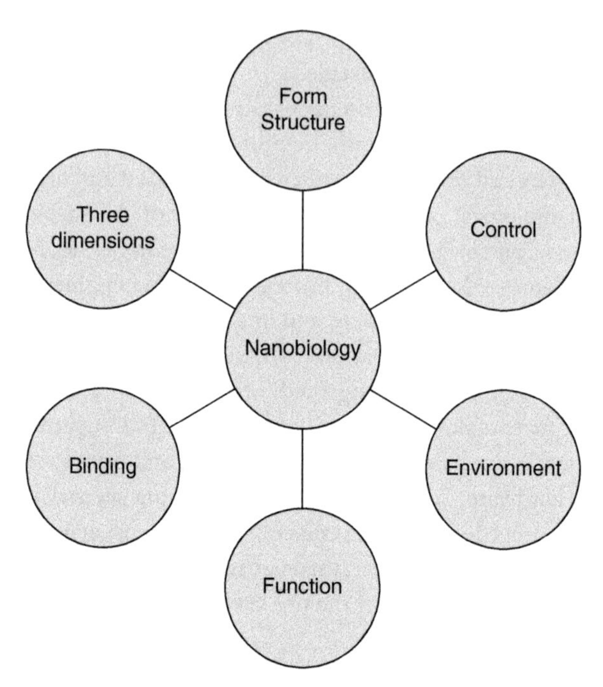

Figure 3.4 The nanobiology hexagon.

(1) *Structure and form* refer to the scaffolding and the architecture of the object on the molecular level, and also to external surface topology and internal structures. (2) *Three-dimensional representations* provide invaluable spatial information on objects, which can then be studied in terms of their volume and the room that they occupy in the space. Volume necessitates taking into account the spatial constraints linked to it. Three-dimensionality also serves as a point of departure for reasoning in which calculations, derived from three-dimensional coordinates, elucidate intrinsic dynamical forces and interactions. (3) *Binding* concerns the capacity of two different biological molecules to connect with one another, or alternatively, for one molecule to change shape as an internal element connects with another. This is referred to as folding. Both categories of binding are constrained by structure and form, and are also directed by forces of different types such as Van der Waals forces, or hydrogen bindings. (4) *Function* refers to these specific actions of a biological substance in connection with the system (be it the set of molecules in which it is integrated, or a more complex unit) in which it operates. Binding and function are affected by the characteristics (electronic, chemical, thermal, hydrometric) of the environment. (5) *Environment* of a biological molecular substance is of foremost importance to the understanding and even control of its form and function. (6) *Control* is achieved through the utilization of particular sets of instruments that often operate in terms of three-dimensionality, and the increasing ability to foresee outcomes obtained through numerical simulation.[34]

[34] The six components of the hexagon have been present in much molecular biology, and particularly in enzymology since the 1950s. We nevertheless contend that the presence of control, as expressed in

Connections between the six parameters are often acute and decisive. It is the linkages and interactions between them that allow us to speak of an integrated nanobiology hexagon that finds expression in much nanoscale investigation. It is not the presence of any single element in the hexagon that proves decisive; it is instead their co-appearance, and notably their collective interactions. For example, protein binding is governed by three-dimensional properties of biomaterials and by the characteristics of the environment, such as thermal values, acidity, humidity, and packing conditions. The combined considerations of three-dimensionality and binding govern the conformation of the molecule and therefore its function. Through controlling molecular conformation, it becomes possible to control bindings that affect function. In this prospect, the combinatorials of instrumentation, described above, find its full interest and expression.

Although questions of function and binding have been explored since the 1930s and even before, and three-dimensional models arose in the 1940s, 1950s, and 1960s,[35] it is in the framework of nanoscale research that all of these elements are combined. The essential ingredient is the capacity to study processes at the molecular level, to "see" the molecules, to predict their configurations and behaviors, and to control outcomes.

Indeed, since the origins of biochemistry in the early twentieth century, proteins have been studied with reference to their chemical composition and function.[36] But the possibility to be able to analyze precisely their internal activity in terms of physical operations, properties, and morphology is recent and overlaps with the birth and rise of nano. The emergence of nanoscale research has promoted the study of relationships between protein function and their active sites. It similarly elucidates some connections between function and environment.

How are the hexagon constitutive elements held together? The thread that performs this cohesive function—that allows the elements to address one another—is energy. Energies in biological molecular systems entail a variety of forces: electronic, thermodynamic, mechanical and kinetic, and magnetic. In biology, matters of energy are particularly expressed with reference to chemical bonds. Energy distribution inside molecules such as proteins, determines the way they are structured, they fold, or they relax. The concept of energy landscape is particularly relevant here. This perspective was introduced by the chemist, Peter G. Wolynes (Department of Chemistry, Rice University) in a seminal article entitled "The Energy Landscapes and Motions of Proteins," published in *Science*

---

bio-nanoscale research, affects the relationships between the different components, and to some measure thus distinguishes nanobiology in interesting ways from neighboring and previous domains of biology and genetics. We thank Professor Francis-André Wolman (CNRS) for his remark on this topic.

[35] M. Morange (1998) *A History of Molecular Biology*. Cambridge, MA: Harvard University Press; P.G. Abir-Am (2006) Molecular biology and its recent historiography. A transnational quest for the "Big Picture", *History of Science*, 44: 95–118; H.-J. Rheinberger (1997) *Towards a History of Epistemic Things: Synthesizing Proteins in the Test Tube*. Stanford, CA: Stanford University Press; E.F. Keller (1990). Physics and the emergence of molecular biology: A history of cognitive and political synergy.

[36] L.E. Kay (1993) *The Molecular Vision of Life*; J.A. Witkowski (2005) *Inside Story: DNA to RNA to protein*. New York, NY: Cold Spring Harbor Laboratory Press; J. Reynolds, C. Tanford (2004) *Nature's Robots: A History of Proteins*. Oxford: Oxford University Press.

in 1991.[37] An energy landscape is a mapping of all possible conformations of a molecule, or the spatial positions of interacting molecules in a system, and their corresponding energy levels. The concept is particularly useful when examining protein studies.[38] The understanding and the description of protein binding and folding often depend on knowledge about the amount and localization of energy present in the molecule. The energy landscape perspective introduces a statistical approach to energetics of protein conformation. The three-dimensional form that a protein adopts, and the pathways that it took as it stabilized in this configuration are determined by the energy distribution. In all of the processes of a particular protein folding, its environment is taken into account in terms of energy. Energy landscape is frequently crucial in research which envisages protein design and control. A last element that indicates the richness of this concept is that it offers the capacity to achieve an understanding of higher levels of complexity in biomolecules, such as the reversible character of their processes.[39]

3.4 NANOBIOLOGY RESEARCH AT WORK: FROM CODES TO CONFIGURATIONS

Nanoscale biology marks an important departure from recent biological investigation. First, the birth of molecular biology has until recently largely been expressed in terms of what Lily Kay convincingly refers to as codes.[40] The genetic code was progressively discovered between 1953 and 1966. This code consists of the arrangement and permutation of four bases: adenine, guanine, cytosine, and thymine (AGCT), which were studied mainly in molecular biology, and particularly in the perspective of deciphering genomes. The results of this research took the form of long streams of letter codes. By contrast, in the perspective of nanobiology, objects are studied in terms of their physical configuration in spatial occupancy. Stated differently, objects are viewed with reference to the relationship between internal structure and surface form. This is made possible by the new generation of metrological devices and computational instrumentation, and by the production of images that are considered by many practitioners to yield a valuable physical rendering of the molecules. This is a move from mathematical logic to a spatial representation. The genetic code is no longer studied for itself, but instead is used as a resource to control the shape of biomolecules—DNA and proteins. This opens a whole field of investigation at the nanoscale on configuration and binding. In this perspective, two areas of research are prominent in nanobiology: one is

[37] H. Frauenfelder, S. Sligar, P.G. Wolynes (1991) "The Energy Landscapes and Motions of Proteins," *Science*, 254:1598–1603. This article has been cited more than 1600 times. <http://chemistry.rice.edu/FacultyDetail.aspx?p = ACC7DC090095C11C> (consulted 3 June 2011).

[38] <http://en.wikipedia.org/wiki/Energy_landscape> (consulted 14 February 2012).

[39] J.D. Bryngelson, J.N. Onuchic, N.D. Socci, P.G.Wolynes (1995) "Funnels, Pathways, and the Energy Landscape of Protein-Folding _ A Synthesis", *Proteins-Structure Function and Genetics*, 21 (3): 167–195.

[40] L.E. Kay (2000) *Who Wrote the Book of Life? A History of the Genetic Code*. Stanford, CA: Stanford University Press.

centered on diverse aspects of proteins; the second entails operations on DNA (DNA nanotechnology).[41]

3.4.1 "Nano-style" proteins: nanoproteinomics

In nanoscale research, the very language of proteins is punctuated with reference to morphology. Examples of this are: configuration, conformation, self-assembly, lock and key,[42] links, loops, close, open, twist, spiral, docking, etc. Conformation and configuration are fundamental descriptors in this domain. An underlying structural component of all of this is encapsulated in what is known as binding. Binding is made visible in the stabilization of the spatial relations between two different inter- or intra-molecular segments. These bonds participate in the building and structuring of a protein.[43] One can summarize the key concept as "adjustment" between two parts of a molecule or between two different molecules, where privileged sites can become target sensors (see Section 6). Figure 3.5 exhibits the complexity of each of the strands composing a protein, the myriad ways in which they can bind one to the other, and the resulting centrality of spatial occupancy.

A protein is a macromolecule that is composed of a sequence of amino acids. Their organization depends on the particular sequence of these amino acids. Proteins can fold in a variety of ways. Binding folding is linked to a host of necessarily accompanying activities that include structure, function, form, and intra- and inter-protein environment. The less conspicuous environment and three-dimensionality are everywhere and transverse. The six elements of our hexagon are expressed in the above-described components and

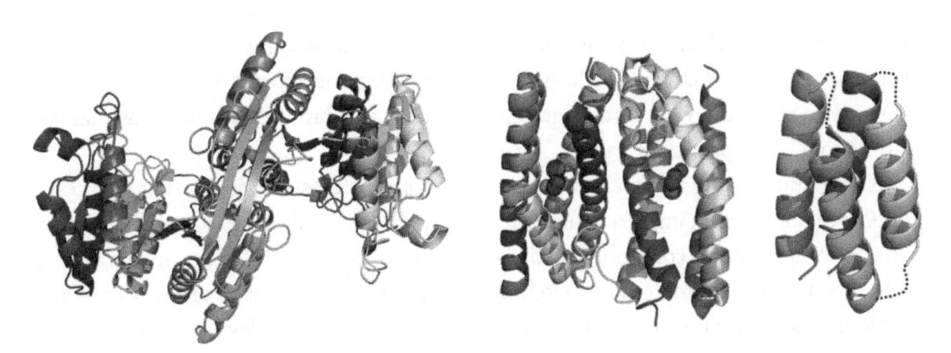

Figure 3.5 Protein binding and folding. Image courtesy of Professor Akihiko Yamagishi.

[41] O. Bueno (2011) When physics and biology meet: The nanoscale case, *Studies in History and Philosophy of Science*, Part C, 42(2): 180–189.

[42] The "lock and key" concept, originally introduced by Emil Fischer in 1894 (1852–1919) (Chemistry Nobel Prize 1902), assumes that one protein has a cavity or indentation that another protein perfectly fits into. Ch.S. Goh, D. Milburn, M. Gerstein (2004) "Conformational Changes Associated with Protein-Protein interactions", Science Direct www.sciencedirect.com. Special Issue on Folding and Binding Edited by D. Baker and W.A. Eaton

[43] Interview of Philippe Minard (IBBM, University of Orsay) by authors, 12 June 2010.

dynamics. Many of these have already been used, individually, in biology, long before the era of NSR. However this has occurred in a fragmented way. This has changed appreciably in the context of nanobiology, where they constitute a system. In this integrated system, the components are interdependent and are viewed in a deterministic perspective. As in physics with the revelation of the selective organization of atoms that dramatically spelled out the letters I.B.M., in nanobiology, the possibility of rendering DNA and proteins visible has opened a new vista on biological materials. Control occupies a pivotal position in the integrated system. We will now describe two research projects on nanoproteinomics. This will allow us to identify the kind of questions asked by nanobiologists and permit us to observe how they conduct their work. The first project deals with binding, the second with architecture, mastering for effects and control.

In this opening example, researchers formulate a complex of questions around issues of protein binding. Practitioners ask how in view of their sequence, a chain of amino acids binds and consequently folds. Alternatively, they explore the forces expressed in processes of binding. Brian Kuhlman, co-laureate with David Baker of the 2004 Feynman Nanotechnology Prize for their work on nanoprotein design, is emblematic of nanoscale research on binding-related nanoproteinomics investigations. David Baker, professor at the University of Washington (Department of Biochemistry), and Brian Kuhlman, professor at the University of North Carolina (Department of Biochemistry and Biophysics), received the theory prize for their development of the above-mentioned Rosetta Design program,[44] which has a high success rate in designing stable protein structures with a specified backbone-folding configuration. A backbone is the primary architecture of a protein that is particularly accessible at the nanoscale. This backbone is a highly complex structure, the understanding of which requires a grasp of innumerable parameters in order to anticipate which form it will take when folding and binding. Binding is what fixes or stabilizes folding. Baker and Kuhlman's goal is to design new forms of proteins. "These proteins we make are very complicated things. They're a hundred amino acids that we want to fold into a particular shape. . ."[45]

Folding and binding are not simply a question of directly sticking together two components of an object. It implies taking into account the encumbrances that must be overcome in order for the components to converge spatially, and this may entail twisting pathways. It is also a problem of matching physical interfaces. This is one aspect of the problem of appropriate atomic and molecular packing that nowadays is probed in NSR. For all of these reasons, binding is observed in terms of forms, and needless to say, as three-dimensional forms. "We have projects where we're just trying to build a certain shape. That's kind of like learning how to build a bridge before I even care about what the bridge is going to connect."[46] Bridging entails control, and it is in this guise that control becomes a centerpiece of Kuhlman's endeavors. Control refers to the mastery of all

[44] As stated in Chapter 2, the Feynman theory prize is awarded for achievements in simulation-based research. <http://www.foresight.org/feynman2004/index.html> (consulted 10 October 2010).

[45] Interview of Brian Kuhlman (University of North Carolina, Chapel Hill) by authors, 18 June 2010. <http://www.foresight.org/FI/2004Feynman.html#2004Winners> (consulted 10 October 2010).

[46] Interview of Brian Kuhlman (University of North Carolina, Chapel Hill) by authors, 18 June 2010.

parameters needed to obtain a form. It is no simple matter to control the complex pathways of a protein that binds to itself. The bonding electronic forces have to be considered and calculated; of equal if not greater importance, the form and size of each amino acid composing the protein constitute an encumbrance in the game of spatial occupancy: bridging the space between point A and point B often involves a highly circuitous route. Through the engineering of amino-acid sequences, nanopractitioners seek to control the structures and forces of spatial occupancy so essential to protein binding.

What work is it necessary to pursue in order to address the question of binding? What is the balance between metrological experimentation and computational simulation experimentation, and what is the relationship between the two? In the case of Kuhlman, metrological experimentation functions as an input that informs simulation. On the whole, much progress is achieved through the latter. Interestingly, in Kuhlman's laboratory, empirical information is employed to correct or enrich software. Software is seen to be learning from the metrological data.

> What we do is we gather all the experimental data that's out there, from all the labs, and we use that to train our software (simulation programs). And if our own experiments aren't working, we then say, 'oh we need to train our software more, with more data that's out there . . . So that's the process we use. I guess, what we use our experimental results for is kind of knocking ourselves on the head and saying, . . . we really need to be doing more training of our software. We need to be doing more testing of our software . . .[47]

Knowledge of nature is acquired through simulations that are trained by data from experimentation and from earlier simulations. It is essential to observe that, in much nanoscience, the underlying objective is to understand nature; but in order to achieve this aim, researchers concentrate their efforts on the construction of artificial molecules. It is the simulation programs that best serve the elucidation of nature in this research style.

In our second case study, another Feynman nanotechnology laureate (2005),[48] Christian Schafmeister, Professor of Biochemistry at Temple University, works on a "universal scaffolding" of artificial proteins synthesized through building-blocks in order to create a functionality of these proteins. In his work two types of control—control as mastering and control for effects—are interlaced. Their simultaneous occurrence in research projects is a signature of nanobiology. Control as mastering refers to the capacity to determine the form, structure, and dynamics of a biological object. In contrast, control of effects concerns manipulation of the output of a biological object—here the focus is function. Control as mastering is associated with learning intrinsic characteristics of biomolecules. As an effect, control is associated with impacting other molecules in the environment.

Schafmeister constructs a universal scaffolding from crystals. In nature, a protein changes its shape when in contact with some other molecule. Structural instability is the habitual condition of proteins under changing conditions. In contrast with this, in Schafmeister's technique, the protein remains stable even when additional components are introduced. It constitutes an architecture, a scaffolding that remains invariant (does

[47] Interview of Brian Kuhlman (University of North Carolina, Chapel Hill) by authors, 18 June 2010.

[48] <http://www.foresight.org/FI/fi_spons.html> (consulted 10 October 2010).

not react) when exogenous molecules are attached. For Schafmeister, form remains central, but it is a means to an end and not an object of investigation per se, as it is in the case of Kuhlman. In other words, he is not seeking to understand form. Schafmeister simply changes form to achieve function. To these invariant scaffolds are attached a number of specific amino acids. Depending on their relative position on the scaffold (this is expressed in three-dimensional space), they generate particular functions such as those induced by acidity or hydrogen bonds.

> Actually, what we're trying to do is, we know what these groups, amino-acid side chains, do by themselves. We know a lot about how they can catalyze reactions by acting as acids, acting as bases, acting as hydrogen-bonding groups. We know how they can coordinate with metals and metals can activate things. What we're trying to do is to put these functions together so they can work simultaneously and act in concert, get a synergy, get an additive greater than the sum of their parts. If you have a group here that's acting as a base, while over here there's a group that's acting as an acid, and over here you have something activating by hydrogen bonding, and they're all held in the right constellation, then when the molecule diffuses in there, it'll be like it's completing a circuit. The electrons will flow, and atoms will move, and out come the products.[49]

This quotation draws attention to the systemic character of Schafmeister's macromolecular scaffolding construction. The functions of the system result from the relations between its components and their environment. The stability he can impose on his constructions gives these systems the possibility of having dynamic interactions with their environment. The scaffold can accommodate a variety of forms, on which also depends the function of the whole molecular construction. In summary, Schafmeister produces control in the sense of mastery by handling molecules; he obtains control in the sense of effects, producing new functionalities, which are created through a triangular system of structure, form, and function.

Finally, what is the relative balance between numerical simulation and metrological investigations in Schafmeister's efforts? For him, an important part of the work consists of writing simulation codes that will allow him to design the proteins and then to adapt them in the light of experimental observations with NMR and X-ray diffraction of the sort described in Section 2.1. Computer graphics are of foremost importance to the simulation work of Schafmeister. A great deal of effort is put into the production of models which are then represented graphically on the computer screen, the idea being to determine scaffolding geometries that best hold things in place. "We're building models of our molecules, and then on the computer, looking for which ones could hold the groups in the right place."[50]

The reader may legitimately ask what is the place of nanobiology in all of this? In Schafmeister's case, the intention is to build universal nanoscale architectures with designed strands of proteins for precise biological functions, which thereby become

[49] Interview of Christian Schafmeister (Temple University, Philadelphia) by authors, 20 June 2010.

[50] Interview of Christian Schafmeister (Temple University, Philadelphia) by authors, 20 June 2010.

the support of said functions and processes, at the molecular level. For Kuhlman, and as he has himself stated, he constructs bridges without knowing where they lead. This research, that may be interpreted as being underpinned by an interest in nanometric forms and their elaboration, privileges biomolecules because of their high complexity and abundant possibilities for morphology. Stated differently, Kuhlman and collaborators are fundamentally interested in elaboration of forms where biological substances offer incomparable potential. In both cases, Kuhlman and Schafmeister, the gap between their work on complex molecule morphology and biological questions, such as the function of these molecules, is filled with their attention to possible applications in biology and medicine.

3.4.2 DNA nanotechnology

A well-defined and moderately large research community conducts investigations in a new domain referred to as DNA nanotechnology.[51] Inside the United States alone, this domain counts over 25 competing/collaborating teams. The very fact that three specialists in DNA nanotechnology have received the Feynman Nanotechnology Prize since 1995—Ned Seeman (New York University, 1995), Paul Rothemund and Eric Winfree (both at Caltech University, 2006)—is a good indication of the standing and stability of this community. In the work on proteins described above, they are studied for their biochemical characteristics, properties, and behavior in vitro and in vivo. In the investigations that we now present, which focus on DNA, the DNA is considered as a means and not as a self-referencing object of investigation. What problematic does DNA nanoresearch entail? What are the features of the object on which the practitioners work? What operations are included in the research activities of this area?

In the past, the unique perspective of studies on DNA has referred to the genetic properties of the codes included within the DNA chains and to the information which they convey. In this way, DNA has been considered in its relationship with biological elements, like RNA, ribosomes, and proteins, which work together in the transmission of information contained within the DNA. DNA has also been studied for the biochemical properties involved in its relations with enzymes, whose role is to open and close the famous DNA double helix in order to read the genetic code.[52] Today the orientation is different. Since the rise of nanofocused biology, DNA is a material used for the construction of molecular objects whose form is of foremost importance. The nanobiologists who work on DNA in this perspective are architects/engineers, who design the forms and then build them.

Ned Seeman is the inventor of DNA nanotechnology. He studied biochemistry at the University of Chicago and crystallography at the University of Pittsburgh. He is most noted for his development of the concept of DNA nanotechnology, dating from the early

[51] O. Bueno (2011) When physics and biology meet: The nanoscale case.

[52] L.E. Kay (2000).Who Wrote the Book of Life?

1980s.[53] Early in his studies, he grasped that biochemistry was not a valuable path—in his view, one is either a biologist or a chemist.[54] Seeman took the path of crystallography. Although specialized in crystallography, he nevertheless continued to interact with biologists, and it was in this context that he asked why should biological materials not be used in the construction of crystal structures. This perspective led him along the way to nanobiology. In the first stage in his evolution toward nanobio, he grasped the possibility of using DNA as a malleable entity applicable to crystal structures. Of foremost importance, in this perspective, he conceived of DNA as a material for producing geometries and not just as codes. Finally, Ned Seeman was a fan of the Dutch artist Maurits Cornelis Escher's (1898–1972) woodcuts. The forms depicted by Escher showed the way to complex geometries that he might be able to engineer on the basis of connected DNA segments.

But why was Escher so interesting for Seeman's work? Escher's woodcuts depict repetitive interlaced structured forms. In his work, one can see a three-dimensional interlocking of complex geometrical shapes. For example, in his 1953 woodcut entitled "Relativity," one observes an arrangement of highly confusing, yet intertwining periodic staircases (see Figure 3.6). At first sight, the image is disorienting, yet it is nevertheless rigorously structured. Seeman's object was to create such complicated geometric structures using DNA. His project was not to understand the structure of DNA, or how it is reproduced or transcribed. For Seeman, DNA is simply a building-block. He did not ask "what" but "how." In this line of research, in 1989, Seeman's laboratory published the synthesis of the first three-dimensional nanoscale object, a cube made of DNA.[55] Seeman could envisage using molecules such as DNA to iteratively build three-dimensional complexes (see Figures 3.6 and 3.7).[56]

During the first decade of his efforts, Seeman's work consisted of the then difficult task of producing quality batches of DNA sequences that would serve as the basic architectural units of his objects. His methodology was to precisely position and maintain the DNA strands using three kinds of connections: the Holliday junction,[57] ligands,[58] and sticky ends.[59] In the early days, Seeman observed his objects using X-ray diffraction; however, with the advent of the nano instrumentation revolution, he incorporated devices such as the AFM, and increasingly used simulation in his efforts. In some respects, Seeman's

[53] J.A. Pelesko (2007) *Self-Assembly: The Science of Things That Put Themselves Together*. New York, NY: Chapman & Hall / CRC.

[54] Interview of Nadrian Seeman (New York University) by Terry Shinn, 28 January 2008.

[55] J.H. Chen, N.R. Kallenbach, N.C. Seeman (1989) A Specific Quadrilateral Synthesized from DNA Branched Junctions. *Journal of The American Chemical Society*, 111(16): 6402–6407.

[56] N.C. Seeman (2004) Nanotechnology and the double helix, *Scientific American*, 290(6): 64–75.

[57] A Holliday junction is a mobile junction between four strands of DNA. The structure is named after Robin Holliday, who proposed it in 1964 to account for a particular type of exchange of genetic information. <http://en.wikipedia.org/wiki/Holliday_junction> (consulted 7 March 2010).

[58] A ligand is an ion or molecule (functional group) that binds to a central metal atom to form a coordination complex. The bonding between metal and ligand generally involves formal donation of one or more of the ligand's electron pairs. <http://en.wikipedia.org/wiki/Ligand> (consulted 30 April 2011).

[59] DNA end or sticky end refers to the properties of the end of a molecule of DNA. <http://en.wikipedia.org/wiki/Sticky_and_blunt_ends> (consulted 6 June 2011).

Figure 3.6 Escher 1953 woodcut "Relativity." M.C. Escher's "Relativity" © 2013 The M.C. Escher Company—The Netherlands. All rights reserved. <www.msescher.com>.

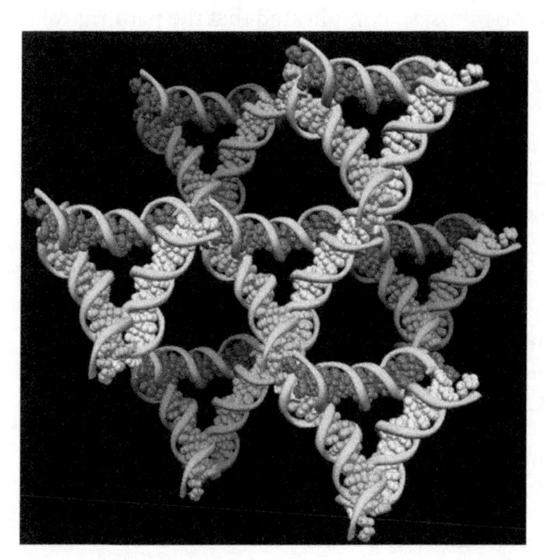

Figure 3.7 Self-assembled three-dimensional periodic DNA crystal reminiscent of Escher's woodcut. Image reproduced with permission of Nadrian Seeman, New York University.

activities depended largely on, and only become intelligible in terms of, the instrumentation combinatorials of nano- and pre-nanodevices which emerged through the 1970s and 1980s.

All of this is part of bionanotechnology and in particular DNA nanotechnology. By using these different trans-instrumentation and biomaterials linking possibilities, Seeman has been able to move from four-sided, to six-sided, to eight-sided, etc. three-dimensional geometries, with increasingly complex relationships. His aim was to create DNA shapes for the purpose of material mastery. This is an instance of structure for the sake of structure and form. Stated differently, Seeman created a control methodology to structure DNA crystals having a pre-designed geometric shape. This signifies playing about with molecules at the nanoscale. Seeman's performance launched a new research orientation that now engages dozens of laboratories and scores of nanopractitioners.

Along the same lines, Paul Rothemund (a former student of Seeman's) has continued this orientation and complexified the geometric forms and enhanced control by designing and manufacturing what he calls "DNA origami." Rothemund is a senior researcher at the California Institute of Technology. His training included a range of domains, from biology to engineering, and extending from applied science to computer science. He received the Feynman Nanotechnology Prize, with Erik Winfree,[60] in 2006 for their work demonstrating that DNA tiles can be designed to form "crystalline nanotubes that exhibit a stiffness greater than the biological protein nanofilament actin, . . . established that algorithmic self-assembly could work well enough to generate non-trivial, non-periodic patterns."[61]

The introduction of DNA origami by Paul Rothemund in 2006 constitutes a development of the DNA geometries earlier established by Seeman.[62] DNA origami introduces a high measure of integration that is now a characteristic of nanobiology.[63] Like the geometries of Seeman, Rothemund's origami forms a base composed of DNA, but unlike Seeman, he mobilizes long strings of DNA extracted from viruses. It is from these extended strings that he composes his complicated origamis; so complicated that the patterns take on the form of geographic maps, snow crystals, or smiley faces . The amazing precision and intricacy of his DNA geometries are shown in Figures 3.8 to 3.10.

> In 2006, I reported a method of creating nanoscale shapes and patterns using DNA. Each of the two smiley faces above, . . . are actually giant DNA complexes imaged with an atomic force microscope. Each is about 100 nanometers across (1/1000th the width of a human hair), 2 nanometers thick, and each is comprised of about 14 000 DNA bases. . . . I call the method "scaffolded DNA origami" . . .[64]

[60] A professor at the California Institute of Technology, in the same departments as Paul Rothemund: Computing and Mathematical Science and Bio-engineering, Computation and Neural Systems.

[61] <http://en.wikipedia.org/wiki/Foresight_Institute_Feynman_Prize_in_Nanotechnology> (consulted 9 February 2011).

[62] Interview of Paul Rothemund (CalTech) by Terry Shinn, 26 January 2008.

[63] P.W.K. Rothemund (2006) Folding DNA to create nanoscale shapes and patterns, *Nature*, 440(7082): 297–302.

[64] <http://www.dna.caltech.edu/~pwkr/> (consulted 3 May 2013).

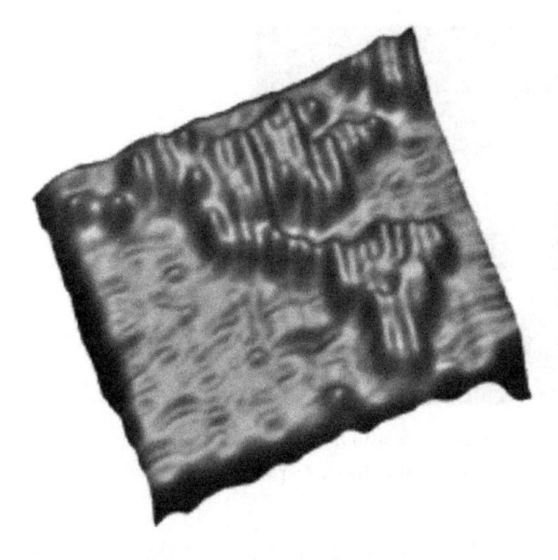

Figure 3.8 A nanometric scale origami map of the Americas constructed from DNA molecules. Image copyright Paul W. K. Rothemund (<www.dna.caltech.edu/~pwkr>).

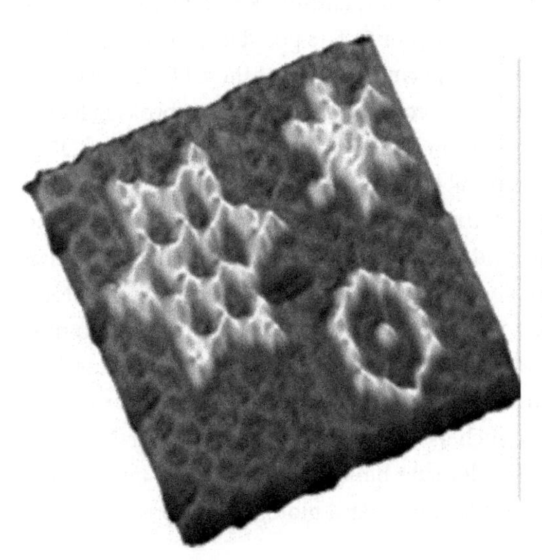

Figure 3.9 A nanometric origami snowflake, constructed from DNA molecules. Image copyright Paul W. K. Rothemund (<www.dna.caltech.edu/~pwkr>).

However, like Seeman, he also uses synthesized short segments of DNA to bind the long strings together. In so doing he mobilizes the biochemistry of the ACTG DNA bases.[65] His origamis entail processes of self-assembly. Perhaps the most fundamental advance obtained by Rothemund is the capacity to integrate non-DNA components into his objects. Proteins can be fixed to DNA in pre-designated patterns. The same can be done with quantum dots. This means that origami may serve as components for more

[65] The DNA bases are: adenine, cytosine, guanine, and thymine. See: <http://en.wikipedia.org/wiki/Base_pair> (consulted 2 November 2013).

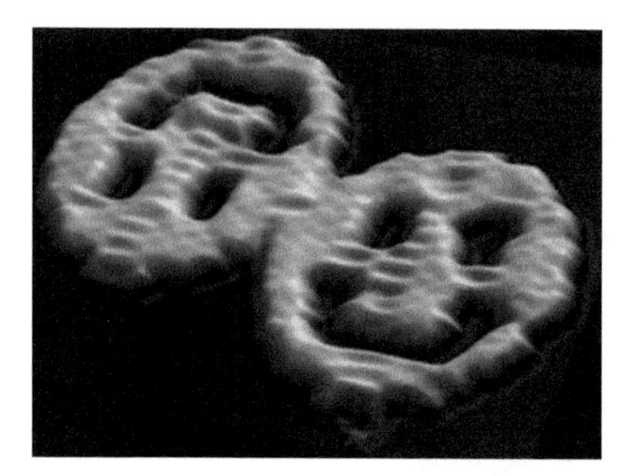

Figure 3.10 Nanometric scale origami faces constructed from DNA molecules. Image copyright Paul W. K. Rothemund and Nick Papadakis (<www.dna.caltech.edu/~pwkr>).

complicated objects. They can be used in nanobio research projects, where they perform the function of markers or detectors. The insertion of quantum dots in DNA molecules is a broader question which will be treated in Chapter 5. The possibility of being able to design, at will, DNA strands with a certain shape permits the use of these DNA origami as templates. Simulation operations are an essential component of research projects focusing on DNA origami. For Rothemund, simulation is the way forward for the use of DNA as a patterning material and for spatio-temporal control.

How does the protein research of Schafmeister and Kuhlman compare with the DNA oriented investigations of Seeman and Rothemund? We indicated that Kuhlman and Schafmeister are indirectly concerned with the types of function that their biological molecules could perform, although it is not their central theme. In contrast to this, Seeman and Rothemund's goal is to produce nanoscale architectures with biomaterials, where the biomaterials are exclusively used as a means of control. DNA is controlled in order to generate pre-designated shapes, such as "templates." In this perspective, their output can be seen as tools independent of all issues biological. In Chapter 4 we discuss the use of DNA tiles of different shapes for additional extra-biological purposes.

3.5 LIFE AS A DYNAMIC LEGO GAME

In this section we introduce a concept that we refer to as "lego." We develop this concept in order to better understand some of the underlying activities present in the work of nanobio practitioners. We suggest that it constitutes an implicit framework that renders their endeavors more intelligible. It is also to be noted that some nanobiologists occasionally themselves employ the term lego when discussing their own work. In this case, it seems to color not only their vocabulary but their very way of thinking; and we can legitimately ask if this is not more broadly the case even among those who do not author the term.

Precisely what do we mean by lego? Lego is a children construction game invented in 1949 and is now ubiquitous, and has even penetrated the world of video computer games. The lego pieces exhibit several characteristics. First of all, in one sense they are universal: each block contains a "lock" surface and a "key" surface and through this, every block can be connected to any other block. Connectibility is universal. The possible form of the blocks is infinite. Clusters of blocks can be assembled and disassembled through processes of attachment and detachment. We will now demonstrate the power of the lego metaphor in a very important area of nanobiology, namely "protein domains."

Protein domains are made of sequences of amino acids which are linked one to the other in a chain (between 25 and 500 amino acids in length). They constitute highly important component of proteins. A protein domain can evolve, function, and exist independently of the rest of the protein chain. It constitutes a compact three-dimensional structure and can often be independently folded in a stable configuration. The concept of a domain was first proposed in 1973 by D. B. Wetlaufer,[66] when he presented his research results on defined stable units of protein structures that he had obtained in isolating pieces of enzyme chains that could fold autonomously (hen enzymes and immunoglobulins). Many proteins consist of several structural domains, and one domain may appear in a variety of different proteins. Because of the specificity of their form and function, and because of the relation with their immediate environment and the possibility of control, a protein domain can be compared with bio-nanoscale building blocks, or lego. Nature often brings together several domains to form multidomain and multifunctional proteins with a vast number of possibilities. In a multidomain protein, each domain may fulfill its own function independently, or in a concerted manner with its neighbors. Domains can serve as modules for building up large assemblies. The simplest multidomain organization seen in proteins is that of a single domain repeated in tandem. The domains may interact with each other (domain–domain interaction) or alternatively remain isolated, like beads on string. The structure of domains is a combination of universality and singularity. In a word, domains can thus be interpreted as selectively transversed structures.

The power of the notion of lego in nanobiology finds further expression in protein domains as applied in the description of the evolution of life. The succession of nano-related research projects revolving around proteins, DNA, architectures, and the combined use of specialized nanometric and numerical instrumentation have progressively allowed some nano-inclined practitioners to move toward the area of directed evolution in a spirit of nano. This connection between lego, domains, and the question of evolution casts light on the research of the nanobiologist Philippe Minard, briefly introduced

[66] D.B. Wetlaufer (1973) "Nucleation, rapid folding, and globular intrachain regions in proteins", *Proceeding of the National Academy of Science USA*, **70** (3): 697–701.

above. Philippe Minard was trained in biochemistry and specialized early in his career in folding and refolding kinetics of proteins. The framework of domains studies has been omnipresent for him. One of Minard's research projects focuses on understanding the functioning of a highly important protein in the economy of the cell.[67] The function of this protein (the RNAse) is to neutralize and eliminate the enzyme ribonuclease, which is both necessary and dangerous: it selectively and intermittently controls the presence of RNA (a nucleic acid essential for gene information transmission) in the cell. The protein (RNAse) which neutralizes the enzyme (ribonuclease) thus plays a fundamental role. The nanobiological interest in the molecule resides in its form, which is particularly intelligible in the lego paradigm.[68] The general shape of the RNAse protein is a horseshoe, a diadem. It has an active inner surface which grasps its targets, and an irregular outer profile that functions as architecture.[69] These characteristics are readily observable in Figure 3.11.

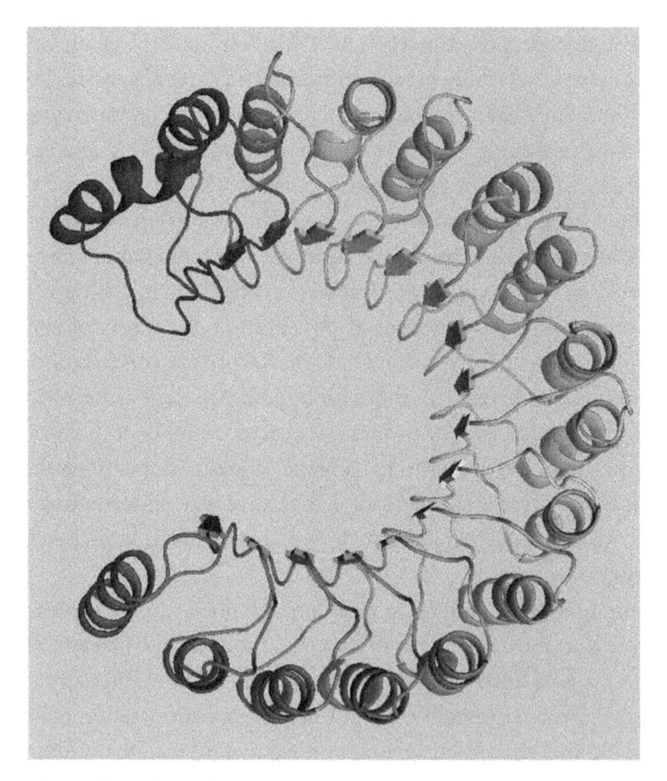

Figure 3.11 A ribonuclease horseshoe. Image reproduced with permission of Philippe Minard, IBBM-University of Orsay.

[67] Interview of Philippe Minard (IBBMC) Université d'Orsay by authors, 12 June 2010.

[68] A. Urvoas, M. Valerio-Lepiniec, P. Minard (2012) Artificial proteins from combinatorials approaches, *Trends in Biotechnology*, (10): 512–520.

[69] email communications from Philippe Minard to authors (3 June 2013, and 3 November 2013).

The diadem consists of a succession of mini-lego molecules. Each of these molecules assembles small springs oriented from the outside of the diadem toward the inside. By isolating, identifying, and understanding each of these elementary components (the springs) and the way they link one to the other, one comprehends three important features: (1) mini-legos assemble to constitute larger legos whose general form is remanent—here a horseshoe.[70] There are myriad horseshoe-shaped proteins in nature, and together they constitute a family.[71] (2) These big lego-shaped proteins present a measure of variability. The variability of the outer end of the springs is not important as they are the architecture of the molecule. Fine gradients of difference in the opposite end, though, have important consequences, as they are the active site of the molecule. As a matter of fact, they nevertheless share significant similarities: the biological function of the horseshoe is to detect and capture specific particles in their environment. This is how the diadem works in the case of RNAse, which is capable of efficiently grasping and neutralizing infinitesimal amounts of ribonuclease in their environment. (3) This lego game emphasizes the tinkering work of nature in evolution. As expressed by the 1965 Nobel laureate in medicine, François Jacob (1920–2013): "Nature is a tinkerer and not an inventor."[72] In the case of the protein domain that we present here, tinkering refers to the modification, addition, or subtraction of the inner tip (the active detecting part of the spring) in response to environmental changes. In nature, a huge number of candidates emerge in the form of the aforementioned tips, some of which succeed—in effect tinkering; as against invention which refers to engineering (intentionality, design, construction) of a calculated solution. Through this understanding of the dynamics of natural selection, scientists can more effectively induce procedures that lead to the development of new artificial protein molecules, through directed evolution. In nanobio research, directed evolution means identification of the lego molecules which work and those which do not.[73]

3.6 BINDING, SENSING, AND DETECTING

In the final section of this chapter, we discuss the massive quantity of combinatorials present in nanobiology and the extent to which they underpin the area. In the bio-phenomena

[70] A. Loksztejn, Z. Scholl, P.E. Marszalek (2012) Atomic force microscopy captures ribosome bound nascent chains, *Chemical Communications*, 48(96): 11727–11729.

[71] This suggests that horseshoe morphology is amenable to addressing many molecular situations occurring in nature. In his 3 November letter, Philippe Minard documents another protein architecture (the alphaReP) consisting of repetitions of the same pattern which then assemble in an infinite range of possibilities and whose slight modifications in structure of individual lego pieces determine variability in the function of the resulting protein.

[72] F. Jacob (1977) "Evolution and tinkering", *Science*, **196**(4295): 1161–6; F. Jacob (1970) *La Logique du Vivant. Une Histoire de l'Hérédité.* Paris: Gallimard; F. Jacob (1987) *La Statue Intérieure.* Odile Jacob éditions.

[73] Interview of David Bensimon, Laboratoire de Physique Statistique, Ecole Normale Supérieure de Paris, by authors, 22 April 2009. One of his research fields revolves around life evolution through the particular problems of DNA–protein binding.

of binding, sensing, and detecting, combinatorials are visible in the conjoining of function and structure, which in turn connects with binding. Another terrain lies in linkage between chemical and electron action. Finally, the combinatorial between metrology and numerical simulation is omnipresent.

In the preceding sections, binding concerned mainly a biological molecule (nucleic acid or protein) which folds and binds two different parts of itself and which grows by self-assembling. In the case of proteins, the function of the molecule is dependent on its shape, and many studies are aimed at constructing specific protein forms in order to obtain determined functions. But binding also implies the possibility for a molecule to connect to another molecule. This is the case in the self-assembling process of biological molecules growth, where a given part of a molecule, here a protein, connects to a privileged site of another molecule. This constitutes the first condition for the conception and fabrication of a biosensor. A biosensor is a nanoscale device for the detection of a substance that combines a biological component with a physicochemical detector. The sensitive biological element, a biologically derived material or biomimic component, interacts with (binds to or recognizes) the target substance. The biologically sensitive elements can also be created by biological engineering. Physical or chemical outputs (optical, piezoelectric, electrochemical, etc.) of a transducer (the detector element) transform the signal resulting from the interaction of the target substance with the biological object into another signal which can be more easily measured and quantified.

The work of Homme Hellinga (Department of Biochemistry, Duke University)[74] is emblematic of this research orientation. He was trained in computer science and molecular biology. This twin training is crucial for understanding the orientation of his research and is emblematic of the combinatorials that have contributed so much to the growth of the bio-nanoscale field. He won the Feynman Nanotechnology Prize in 2004 for his work on the computational design and production of a protein.[75] This protein had a specific binding capability enabling it to attach a small target molecule. This combination of computational design of a protein with a biosensor problematic contributes to the originality of his work.

One of the first tasks of Hellinga's team revolved around the redox reaction and on natural redox centers in proteins (redox—short for reduction–oxidation—is the chemical process of electron transfer), which figures importantly in the generation and reception of signals that are so central to detection and hence to sensors. This is an example of a combinatorial associating the design of a protein site and the control of electron transfer by the inclusion of an iron cluster.

> What we were doing is an iron–sulfur center as a natural redox center. You see this in many different proteins. Their function is essentially to absorb and desorb electrons . . . The flow of electrons in biological systems is what life is all about. . . . But we asked, can we create a

[74] <http://www.biochem.duke.edu/modules/biochem_hellinga_lab/index.php?id = 1> (consulted 3 October 2012).

[75] <http://www.foresight.org/FI/StudentAward2004.html> (consulted 9 July 2010).

protein that can have an iron–sulfur cluster inside of it? And will that ion–sulfur cluster then have the appropriate electron absorption–desorption redox properties?[76]

We see here the construction of a combinatorial: (1) the understanding of the redox activity of a specific protein; (2) the idea of transforming this redox property into another redox activity (changing the iron site into an iron–sulfur site), in order to make this site able to emit distinguishable signals (different colors for the donor and receptor); (3) the work of making a simulation program to design the protein in order to make it bind the redox cluster; (4) the production of such a protein. The question here is to reproduce and mimic life processes at the nanoscale and to control and manipulate them in order to give this biological material a specific function—binding, sensing, and detecting.

Another element in Hellinga's work is its complete reliance on the property of self-assembly in biological molecules as they bind to one another. Here the issue is complementarity of forms. The shape of each object must be such that combination is possible. The interfaces of each molecule have to fit one to the other. So as a "designer" at the nanoscale, Hellinga essentially shapes the components of a protein so that they self-assemble into the intended structure. The underlying logic is combinatorial.

It's a little bit like—a good analogy at one level is making a jigsaw puzzle. As designers we are drawing the shape of each puzzle piece. And when they fit together appropriately, that's what you would call an assembly. Unlike a jigsaw puzzle, where we are doing the assembly, in nature, it's the fundamental inter-molecular forces that tend to assemble things together, and disassembling things, and by assembling and disassembling all the time, you eventually come up with a puzzle piece that fits correctly and the whole puzzle starts to form. . . . change the shape of the interface between the two proteins, by changing the sequence of one of them, I will change how the shape will assemble, or whether it assembles at all. That's one example . . . The cavity of the protein that binds to small molecules is, in technical terms, shaped in a complementary way . . . But that's the essence of structure. A structure is what it's all about. Hence my emphasis of structure-based design.[77]

This combinatorial approach, which privileges sensing and detection, starts with a protein of known structure and uses simulation algorithms to predict the set of mutations necessary to alter or introduce ligand binding or enzyme activity in that structure. This computational work results in a series of predictions, that is a series of mutant proteins, which are then produced in the laboratory. As Hellinga explains:

We were able to design proteins with pre-specified functions and particular proteins that were able to act as receptors, in fact, to change the sequence of a protein to change its specificity, in order to make them able to bind to small molecules of choice.

This research orientation implies a constant back-and-forth movement between computational design and validation by metrological experiments on proteins, with spectroscopy and AFM. To create a protein whose property is to bind to very specific

[76] Interview of Homme Hellinga (Duke University, Durham, NC) by authors, June 2010.

[77] Interview of H. Hellinga (Duke University, Durham, NC) by authors, 9 June 2010.

small molecules or nanoscale substances (to sense and detect) implies then a very precise knowledge of the relationships between a given structure and a function, and a detailed understanding of the relationships between its shape and its binding properties. This is particularly emblematic of nanobiology. An example of such research is the work that Hellinga and his team conducted on maltose.[78]

The idea was to design a protein with pre-specified functions, thus enabling it to act as a sensor of a given substance (here maltose).[79] Using advanced computer systems, Hellinga and his group studied the protein to see how its shape changes when maltose attaches to it. The researchers were looking for a special spot on the molecular surface of the protein where they could bind the reporter molecule (a molecule that changes its emitted fluorescent light when the shape of its immediate surroundings is altered). Here, fluorescence signals the presence of maltose. The molecular perspective and molecule to molecule interaction is typical of nanoscale research, be it in biology or the physical sciences. The trick, said Hellinga, was to locate the spot where the reporter feels the movement in the structure of the protein when maltose becomes attached to it. Using its computers, the team created a special sequence of events: the protein binds to maltose and starts to change. These structural changes affect the fluorophore, which then itself, changes. It emits light, thus providing the desired signal of these molecular events.

The method employed for this production of nanobiosensors includes steps that summarize the ensemble of the bio-nanoscale research themes that we have evoked in this chapter, and show just how important control is in nanoscale research. It is interesting to note also that this domain of investigation mobilizes a concatenation of combinatorials that include: instruments and technologies extending from computational design for shaping the site cluster in the protein to become the sensor, in vivo production of protein sequences, metrological experiments mainly with spectroscopy and AFM, transformation of the protein site by adding devices to make it signal the substances that they are intended to detect as a target in their environment.

CONCLUSION

The significance of the relations denoted in the six features (structure/form, binding, three-dimensionality, environment, function, control) presented in the nanobiology hexagon lies in the links between them, which are exceptionally integrated. Each feature explains and reinforces the others. This interaction stresses the fact that the field addresses highly complex molecules and their particular biological properties. At the same time the interaction points the way to control.

[78] J.D. Dattelbaum, L.L. Looger, D.E. Benson, K.M. Sali, R.B. Thompson, H. Hellinga (2005), Analysis of allosteric signal transduction mechanisms in an engineered fluorescent maltose biosensor, *Protein Science*, 284–291.

[79] Maltose belongs to the family of sugars.

The above study of nanobiology pays attention to two opposite currents: (1) the idea of modularity, which highlights the place of building blocks, their potential for assembly, and their capacity to be engineered at the nanoscale into components; (2) the extreme complexity of molecular structure and dynamics which are affected by myriad variables, the most emblematic of which is the energy landscape. Nanoscale research is characterized by the fact that practitioners straddle both of these currents. Both currents emphasize the centrality of binding. Binding is the confluence of configuration and molecular function. A second common terrain is molecular self-assembly. The third shared orientation revolves around the ambition of nanobio molecular control. Binding, self-assembly, and control are expressed differently in each of the two currents—"modularity" and "energy landscape."

In the case of modularity, the problem is to understand the functioning of a biomolecule, given its configuration and its specific affinities; the effectiveness of the molecule's function is rooted in an architecture that is itself best understood as a nano module. Modules' components as well as inter-modular binding are governed by self-assembling dynamics. One important focus of self-assembling is the variability of components. As we have seen above, this research orientation can open perspectives in evolutionary mechanics at the molecular level. Advanced knowledge about the relationship between binding, the modularity of form, and function enable scientists working in this current to conceive and construct precisely controlled artificial modulable biomolecules.

In the second nanobio research current that we have called the "energy landscape," the question of binding revolves around pathways, occupancy of space, and intramolecular environment in terms of energy. This entails issues of encumberment and energy distribution. These matters emphasize the essential aspect of how things bind. Here issues of form are omnipresent—for example form of protein trajectory and the geometry of the two protein sites that link, as a protein connects to itself or to another. The energy landscape is decisive here. In this research current, the complexity of binding is extremely substantial. The powerful combinatorial of simulation and metrology is sometimes frustrated and the inventiveness of the human mind is mobilized in the guise of public participant protein folding initiatives (Foldit). Here too, self-assembly is a key issue. One particular expression comes to the fore: the origami research programs. Using self-assembly properties of DNA, researchers can program complex attractive interactions into and between DNA components. They can thus generate control over the sequence in which the elements self-assemble. In the energy landscape current, control of bio molecules particularly focuses on the forces which govern three-dimensional molecular shape, the relations between molecules and between them and their environment.

Very little has been said in this chapter about the technological applications of nanobiology.[80] These are a topic of huge public interest and hope, government and private

[80] F. Jotterand (ed.) (2008) *Emerging Conceptual, Ethical and Policy Issues in Bionanotechnology*. Dordrecht: Springer.

investment, and some laboratory research, particularly in those areas related to health.[81] On their websites, numerous scientists state their interest in medical applications such as cancer and the difficulty in targeting tumor cells. To what extent this is linked to hype, to institutional or financial strategies, or to short-term or long-term rigorous laboratory investigation remains to be seen. What is clear is that the topic of application of nanobiology research certainly warrants serious empirical scholarly research by historians and sociologists.

[81] For example P.T.F. Harper, J.D. Wong, S.S. Lieber, C.M. Lansbury (1999) Assembly of A beta amyloid protofibrils: An in vitro model for a possible early event in Alzheimer's disease, *Biochemistry*, 38(28): 8972–8980; J.R. Heath, M.E. Davis, L. Hood (2009) Nanomedicine targets cancer, *Scientific American*, 300 (2): 44–51.

4

Epistemological Frames and Practices

This chapter explores the epistemological referents entailed in many of the processes of knowledge generation in NSR. We identify five epistemological elements: form, image, descriptivism, local, and determinism. The first section, "The anatomy of form," identifies the key unit of observation, information, and reflection in NSR around which description and analyses are framed. As repeatedly indicated in this book, scientists perceive, analyze, and communicate the nano world mainly in terms of the form, size, and relative position of single atoms and molecules, and clusters of these entities. These parameters are perceived as constitutive of structures. NSR practitioners also strongly insist that they often determine the presence and the operation of numerous physical properties. Deterministic reasoning with reference to these dimensions (form and size) contributes to the distinctiveness of nano-epistemology versus the statistical, probabilistic reasoning of quantum physics.

Visual information and representations as building blocks for describing and understanding the world of nanoscopic entities are explored in the second section, "Visual images, form, and knowledge." The epistemology of image incorporates two facets. Firstly, metrological and numerical experiments are designed and conducted with the express expectation of producing images. Secondly, images are explicitly accepted as constituting privileged data necessary to cognition. They are appreciated for providing key information about form and relative positions of objects. Among nano practitioners, the spatial relations depicted in images are held as representing the spatial relations of physical entities. Images enable a dual view of objects: an image can simultaneously permit practitioners exploring a system to study single objects and their form in detail, and at the same time to view the ensemble of objects constitutive of the whole system. They thus promote a synthetic (holistic) and a unitary (segmented) perspective, where scientists may switch easily and rapidly between the two. Implicit in the reflection and communication of researchers is the idea that images are more than information, that they do have a rough correspondence with what lies beyond their detection devices. The centrality of imagery in NSR contributes importantly to the omnipresence of descriptivism in the field.

The varieties of epistemological structures and dynamics occurring in the multiple relationships that exist between research questions on the one hand, and research materials/tailored research-objects on the other is discussed in the third section, "The question to material to material to question spiral." In one scenario, the repertory of legitimate scientific questions is affected by what is accessible and relevant in the catalogue of novel extant nanostructured materials. Symmetrically, due to the current capacity in nanoscience to generate novel substances at will, nanoscientists can now formulate their questions and then request the materialization of appropriate objects. In a third scenario, existing knowledge about a material privileges the framing of certain categories of well-defined questions that require novel forms of control to fully address the problem. This entails supplementary knowledge of the initial substance and considerable capacity of control. Finally, some unusual epistemological reflection focuses on extending what is perceived as adequate knowledge of nanostructured matter—the "materiality of matter." Stated directly, in NSR, materials today become a key epistemological referent. Issues of linkage between interrogation and substance indeed constitute a crucial resource in the NSR epistemological arsenal.

The fourth section, "The epistemology of nanobiology recognition," discusses a longstanding concept in biology, recognition, which in NSR acquires new dimensions and new meanings. It first injects considerations of dynamics into the epistemology of the field. Dynamics refers to action between two entities, one being the target of the other. Dynamics here implies affinity of forms between these entities, occupancy of space, binding, and movement. Processes of recognition at the nanoscale thus illuminate the question of form as selective action.

The chapter closes with "Conclusion: theory, models, and description," with the observation that the role of theory and even of many categories of models, including mediated models, appears to be relatively limited in NSR. Theory as a generalizing, unifying, or law-based system is not to be seen. Prediction is limited to numerical simulations results and arises in the framework of single experiments or highly specific physical events. To the extent that models occur at all, it is in the form of what we term "local models." As opposed to models and theory, in NSR description is paramount and situated in a local setting, and is rooted in strongly spatial considerations. Finally, NSR descriptivism necessarily entails determinism. In summary, the perspective adopted throughout this book, and particularly in this chapter, is a descriptivist one which is shouldered by the epistemologies of the local and of determinism.

4.1 THE ANATOMY OF FORM

Here we discuss the centrality of "form" in NSR. As indicated in previous chapters, in NSR, objects are detected and represented in terms of their form, where form often stands at the heart of cognition. Form-based cognition is not new. The history of knowledge has long been punctuated by morphology. There are innumerable examples of this: astronomical constellations have since Antiquity been described with reference to the form of galaxies, the distances between them, and now their dilation or contraction.[1] Stretching back to the ancients, cartography has been based on the contours of landmasses and relative distances between them.[2] The morphology of animals served as the basis of Carl Linnaeus' the Comte de Buffon's eighteenth-century categories of animal species.[3] In the early twentieth century, Wentworth d'Arcy Thompson explored the relationships between growth processes and the shape of living organisms.[4] Form has long constituted a key referent for distinguishing between different crystals and understanding them.[5] Not least of all, George Gamow presented his 1930s alternative model of the atom's nucleus in terms of form—the shape of a liquid droplet.[6] Form thus stands as one constant in the history of learning.[7] Its importance to research and epistemological power are today further enhanced in many of the investigations carried out in nanoscale research.

Twenty years ago, the nanoscale referred to objects smaller than 100 nm. More recently, the scale has continually diminished, and now almost all nanoscale research is performed on objects of only 30 to 40 nm, or even far smaller, extending down even to 1 or 2 nm. In the physical world, size affects properties, and nano-objects exhibit very different properties from bulk substances. The spatial relationship between surface and mass is radically different in nano. In NSR, it is thus size and form that count, and properties are understood in relation to form. Size, and its expression as form, provides a formal definition of nano. For an understanding of what is studied in NSR, how it is studied, and how epistemology operates, it is necessary to reason in terms of spatial relations. Most research in nano refers directly to the form of objects. Form may allude to the morphology of a particle or to the relations between particles. It is our thesis that form constitutes a principal object of investigation, observation, and understanding.

In NSR form assumes many guises. In each of them, it underpins and imposes a range of epistemological referents and operations. Here we introduce three expressions of

[1] E.P. Hubble (1936a) Extra-galactic nebulae, *Contributions from the Mount Wilson Observatory/Carnegie Institution of Washington*, 324: 1–49; E.P. Hubble (1936b) *The Realm of the Nebulae*. New Haven, CN: Yale University Press; G. De Vaucouleurs (1959) Classification and morphology of external galaxies, *Handbuch der Physik*, 53: 275.

[2] JB. Harley, D. Woodward (eds.) (1987–98) *The History of Cartography*, Volume 2, Book 1: *Cartography in the Traditional Islamic and South Asian Societies*. Chicago, IL/London: University of Chicago Press (2 volumes).

[3] C. Linnaeus (1758) *Systema naturae*. Stockholm: Laurentius Salvius; G. de Buffon (1749) *Histoire naturelle, générale et particulière, avec la description du Cabinet du Roy.* 3 Volumes Tome I Texte établi, introduit et annoté par S. Schmitt avec la collaboration de C. Crémière. Paris: Honoré Champion, (2007–2009); P. Galison, L. Daston (2007) *Objectivity*. New York: Zone Books.

[4] W. d'Arcy Thompson (1907) *On Growth and Form*. Cambridge: Cambridge University Press.

[5] G. Wulff (1901) Zur Frage der Geschwindigkeit des Wachstums und der Auflösung von Krystallflächen, *Zeitschrift für Krystallographie und Mineralogie*, 34(5/6): 449–530; L. Hoddeson, E. Braun, J. Teichmann, S. Weart (eds.) (1992) *Out of the Crystal Maze: Chapters from the history of solid state physics*. New York, NY: Oxford University Press.

[6] G. Gamow (2002) *The Cambridge Dictionary of Scientists*. Cambridge: Cambridge University Press (2nd edition).

[7] E. Dekker (2012). *Illustrating the Phaenomena: Celestial Cartography in Antiquity and the Middle Ages*. Oxford: Oxford University Press.

form that frequently structure and animate research. The categories of form now briefly presented are: (1) form as morphology and landscape, (2) form and the occupancy of space, (3) form, force, and perturbations. In all of these expressions, form as spatial relations comprises the basis of research observation and reflection. Finally, in NSR, form is synonymous with three-dimensional representation.

4.1.1 Morphology

In common parlance, "morphology" is a general term that primarily alludes to shape. But in certain NSR endeavors, the notion of "morphology" entails a more exacting definition. Morphology-based thinking, analysis, and discussion occurring in nanoscience sometimes frame the observation of objects in terms of regular geometries—squares, triangles, hexagons, etc. However, most of the worlds visited by practitioners reveal broken, highly irregular, and even unstable morphologies. Here, the forms are considered as "topologies." As documented in Chapter 2, form is interpreted as a landscape having map-like qualities—mountains, plains, valleys, terraces, etc.[8] Language associated with construction of buildings may be employed. How is form as morphology connected to the objects of research? Morphology refers here to the relative spatial relations of single atoms or molecules, or in the case of a molecule, to its surface geography. The nanoscientist associates shape with physical structure—architecture of objects. Form in the guise of tilting deformation, for example, is sometimes translated into inference about the presence of forces. The evolution of a morphology over time may indicate to the nanoscientist processes of structural composition.

Re-examination of a surface physics research project, set forth in Chapter 2, illustrates the epistemological operations of form expressed in terms of morphology as exercised in some nanophysics investigations. In nanoscale surface-science research, surfaces are often presented as a troubled landscape consisting of high mountains, deep valleys, ridges, peaks, plateaus, etc. These asperities constitute the very objects of research. This orientation that accentuates the place of form indicates the extent to which nanoresearch is descriptive. In thin-layer surface physics, these peaks and valleys are studied as angles, edges, surfaces, descending terraces. These geometries correspond with the positions of molecules or atoms and indicate their possible interactions. The careful study of these forms can inform scientists about electronic and magnetic properties, and about possible transformations and reactions with the environment.

For example, the process of metal oxides in water comprises an important research theme. In Chapter 2, in a discussion about relations between metrological and simulation experimentations in nanoscale research, we referred to the importance of shapes in the explanation of processes. In the study of the dissolution of oxide crystals in water, the main question is to understand how atoms of the oxide surface, progressively detach, and why some do so before others. Observation of the form of the "terraces" and the

[8] Interview of Jacques Jupille (INSP, Paris) by authors, 17 September 2008.

edges of this crystal supplies information that constitutes the language of discussion and reflection, and which finally represents the core of knowledge. Changes of form correspond with an evolution in dissolution. Here again, form emerges as knowledge; in the same perspective, form is a gateway to procuring information on electronic states that accompany this process.[9] The lexicon of these progressive morphologies extends from the cube, with its edges and corners, to the truncation of the edges, to a tetrahedron, to a crenellated form resembling a factory roof. Scientists anticipated an evenly curved form. Discovery of the spiky curves constituting a crenellation was accounted for by reference to an electronic state of the surface. The different morphologies here described express a local landscape. By "local," we mean a defined geography where an event is intelligible with reference to the event's own scale. In NSR, the scale of a phenomenon, intelligibility, is about 1 nm to about 40 nm. Medium and longer distance entities and effects are not included in the frame.

Finally, the concept of "lego" developed in Chapter 3 to describe important aspects of nanobiology equally extends to the epistemology of morphology as form. Lego refers to the constitution of objects from interconnecting building blocks. Entities are the addition of many interfacing components, each of which can be clearly identified, and whose linkages can be specified. This notion of module is current in NSR, where it underpins form in the guise of architecture. Lego and module further figure in morphology as form, since they introduce a parameter of dynamics through the actions of connecting and disconnecting of constitutive components of a larger geometric whole. By the same logic, the dynamic of lego and module constitution and dissolution inject considerations of temporality into morphology.

4.1.2 Occupancy

Occupancy refers to the organization of multiple objects within a confined space. NSR scientists turn to form to explore the features of the elements concerned, how they are arranged spatially, and sometimes the force relations between elements. What geometries occur in the organization of occupancy? Does it consist of complex intertwinings or instead, evenly distributed entities? Is encumberment present, and what are the format and physical issues at hand? Is occupancy chaotic or alternatively laid out? A lexicon of such descriptions identifies and classifies objects, where specificity can relate to function and can provide components of explanation. The epistemological strength of an appreciation of form in the guise of occupancy is that, to an extent perhaps greater than other expressions of form, the powers of structural analysis are particularly pronounced. The fundamental unit of observation, description, and often learning is once again highly descriptive. To describe is to know, or at least, to know much. As in the case of morphology, the landscape is local.

[9] G. Wulff (1901) Zur Frage der Geschwindigkeit des Wachstums und der Auflösung von Krystallflächen, *Zeitschrift für Kristallographie*, 34: 449–530.

Nanobiology investigation affords a complex and insightful demonstration of the operation of occupancy as one expression of form. As we saw in Chapter 3, the question of folding biological molecules, like proteins and DNA, is paramount. Folding is the process by which segments of a molecule change position (move) in order to bind with another segment of the same or an alternative molecule. This displacement implies the problem of spatial constraints due to the occupancy by each component present in the volume, where each of these elements possesses a size and a shape. This is a three-dimensional puzzle, where parts can either open or obstruct the path for other elements to bind. The issue of occupancy is so complex that it sometimes frustrates efforts in simulation and experimentation. As indicated in Chapter 3, nanobiologists have invented an online protein folding puzzle which invites interested individuals from the broad Internet public to mobilize their intuition and visual skills in attempts to identify novel occupancy scenarios related to protein-binding pathways.

Issues of occupancy introduce to the epistemology of form a binary logic. For a specified volume the co-presence of elements is either possible or impossible. At first sight, a folded protein resembles a plate of spaghetti. For the nanoscientist, however, the rigors and rules of occupancy render the protein comprehensible as a spatially organized object. Occupancy-inspired reflection is an important epistemological factor in distinguishing chaos from complexity.[10]

4.1.3 Form, perturbation, and force

Transformations in features of form may register a change in intensity of forces. Stated differently, with reference to a form, at time t_1 and point p_1, the application of a force modifies the aspect of the form at time t_2 and point p_2, through perturbations.[11] Changes in the object's states reveal more detailed knowledge about it. Transformations of shape constitute a kind of probe.

This triangulation between form, perturbation, and force connects form to the epistemologies of change and, by dint of this, to consideration of cause and explanation. This perspective contrasts somewhat with the more static appreciation of form expressed in morphology and occupancy. In line with this, in the perspective of form, perturbation, and force, practitioners' search and reflection move beyond description. Here nanoscientists must define what figures as "distortion" of entities, and in exactly what way the structure of a form proves susceptible to a force; and, not least of all, how the force reshapes the form. In form, perturbation, and force, we identify the operation of an epistemological chain; and indeed something more than a chain, since the chain functions as a dynamic concatenation. It is this expression of form that best identifies processes in the substances explored in NSR.

[10] Interview of Philippe Minard (IBBM, University of Orsay) by authors, 12 June 2010.

[11] A. Marcovich, T. Shinn (2011c) Estrutura e funçao das imagens na ciência e na arte: entre a sintese e o holismo da forma, da força e da perturbaçao, *Scientiae Studia*, 9(2): 229–267.

Finally, our above description and reflection on categories of forms and their place in nanoscale scientific research call for consideration of the relationship between form/ descriptivism and explanation. One can envisage three scenarios: (1) A maturity scenario: Here, NSR would be judged a young science, temporarily intoxicated with the observational possibilities of its innovative instrumentation, which privilege form and drive out alternative epistemological terms, such as explanation. In this view, NSR is in a temporary historical phase that will in the future evolve, permitting the penetration of explanation with an accompanying diminution in the centrality of form. The outcome might either be the peripheralization of form or a balance between explanation and form. (2) An exclusionary scenario. In this view, the domination of knowledge by form, would exclude explanation. This is because form can be regarded as "by its nature" descriptive, and explanation "by its nature" theoretical. This would imply an extreme incompatibility between description and explanation. This scenario, based on irreducible duality, would clearly paralyze deeper reflection. (3) An inclusionary scenario. Form and its accompanying descriptivism can be considered in themselves as explanation. Here, explanation occurs within the framework of form with reference to the exacting precision and the integration of components of an object as given in NSR. This form-based representation elucidates both the architecture and the functioning of nanostructured entities. This elucidation may be seen as constituting an expression of explanation. In this scenario, explanation is an integral component of form. This is tantamount to form as explanation.

4.2 VISUAL IMAGES, FORM, AND KNOWLEDGE

There exists an abundant literature on the production and functions of images in nanoscience and technology. NSR probably hosts more images than most other research fields, which is in part explained by the almost simultaneous emergence and expansion of NSR experimentation technologies and artificial materials, on the one hand, and the introduction of a variety of high-performance imaging devices, on the other. The attention given to images in nanoscience and technology similarly reflects the breadth of the public's (enterprise, governments, and the lay public) interest in the field and thirst for news. Nano has secured a place in the minds of many audiences. Graphic artists increasingly seek to devise images that will give heterogeneous groups some notion of how a nano-object appears, how it functions, and what a nano-based world might look like. Such images are mobilized by business to promote merchandise, or by government agencies to make policy argumentation more convincing.[12]

Not least of all, the existence of a huge and multifarious audience for NSR images increasingly encourages the penetration of esthetics or of special effects in nano portrayals. Here, the aim is to make images more attractive, amusing, dramatic—in effect, to

[12] M. Ruivenkamp (2011) Circulating Images of Nanotechnology, doctoral dissertation, Universiteit Twente.

transform them into consumables. Artists today generate nano-related images. In some instances, practicing scientists too will either simplify their images or introduce dramatic color to images to make them attractive to fellow practitioners and also to extra-science groups. Nevertheless, the horizons of esthetics and the circulation of images among massive and diverse audiences often acutely diverge with the exigencies and logic of cognition, where the relevant parameter is instead the contribution of a NSR image to the stock of knowledge about the physical world.[13]

4.2.1 Images as knowledge

On an analytic register, the images of nanoscience and technology are generated and explored with reference to their depictions of the physical objects of research. However if the question is asked to what degree do the entities rendered in an image correspond with the physical world, no solid reply can be forthcoming. Nano-images are particularly problematic because of the scale of the research entities—a mere billionth of a meter! Hence the images cannot be metrologically verified through comparison with any independent material referent, such as images of macromolecules. Thus in some quarters, one can observe the existence of a legitimate unease about the validity of images of nano-objects.[14] In the language of the sociologists of science and technology, Arie Rip

[13] D.S. Goodsell (2003) Looking at molecules: An essay on art and science, *ChemBioChem*, 4: 1293–1298; T.W. Staley (2008) The coding of technical images of nanospace: Analogy, disanalogy, and the asymmetry of worlds, *Techne*, 12(1): 1–22.

[14] The cognitive, epistemological, and social operation of "images" in nanoscience and technology, even including what legitimately counts as an "image," is a huge, complex, and sometimes heatedly debated topic. Throughout this book we have repeatedly indicated that our interest lies in the collection and development of information in research on nanoscale objects and associated properties and processes, and we contend that in nanoscale scientific research, information often takes the form of images. Some sociologists of science and technology have carefully explored images in NSR produced by artists, and sometimes by scientists in programs animated by imaginings about some projected future. Here depictions serve as vehicles for broad audience public communication, where the images are only loosely linked to carefully coded, technical renderings. See M. Ruivenkamp (2011) Circulating Images of Nanotechnology; M. Ruivenkamp, A. Rip (2010) Visualizing the invisible nanoscale. Study of visualization practices in nanotechnology community of practice, *Science Studies*, 23(1): 3–36. This orientation may be regarded as part of a more general cultural drift, where images multiply throughout society, becoming omnipresent. Another kind of "imagery" revolves around virtual reality (VR). Such images can sometimes even generate haptic effects, as in the case of the VR representations engineered by the University of Santa Barbara nanophysicist, Paul Hansma, in which an observer can even sense the resistance of biological nanomolecules as they are torn apart. C. Mody has pointed to the many species of images (including Virtual Reality) that pervade depictions associated with NSR. See C. Mody (2004) Small but determined: Technological determinism in nanoscience, *Hyle-International Journal of Philosophy of Chemistry*, 10(2): 99–128. Some philosophers of science have reflected critically and systematically on the issue of the structure and epistemology of images in the domains of nanoscale scientific research. In certain philosophical circles it is argued that the physical attributes of nanoscale objects and properties are sufficiently like the macroscopic world that, to some extent and in some special ways, nano-images are intelligible to human experience grounded in the macroscopic. According to this view, images incorporating well-defined codes and technical characteristics provide a valid window between phenomenology and physics. See T.W. Staley (2008) The coding of technical images of nanospace. Such renderings (technological images) contrast with the un-explicit and dramatized nanoworld

and Martin Ruivenkamp, the danger of what they term the "visualization regress" looms. Their argument is framed with concern for outside, stable data and comparison for establishing image validity. What in a nano-image is artifact or only background noise versus the intended physical object? Nano-images are new, and we lack a substantial and historical stock of material and experience to provide a standard against which images can be measured and critiqued. From these perspectives, some research images may be viewed as more or less problematic.[15]

In the pages that follow, images are explored in terms of the roles they play today in the work of NSR practitioners as they conduct research in the laboratory. The purpose of the image is to provide information about an object under investigation. An image is relevant as a vehicle of data—normally as related to form and position.[16] A good image is one that is informationally rich. The key concept here is "image intelligibility," which alludes to the capacity of the rendering to address the research matter at hand. Stated differently, in this subsection our focus is on how scientists produce images, what they do to them in order to identify and extract wanted information, and what practitioners perceive in images as promoting cognitive intelligibility.[17]

A significant portion of the human brain is given over to visual information and processing of that information.[18] Vision occupies a privileged place in the activities of learning. The recent introduction of improved visual images, and in particular the capacity for three-dimensional representation, are important components in modern science. Seeing and related activities are central to epistemology in NSR. Seeing is even more important by dint of the above-discussed centrality of form in the description and analysis of the nano world. In NSR, seeing entails the development of images that require specific technical operations. We are not concerned with the status of what is perceived in images. Claims about "likeness" between the ontology of the physical world and depiction in images are not relevant to our reflection. We instead gravitate toward the position of the philosopher of science, Otavio Bueno, who suggests that in nanoscience useful

representations manufactured by physicists and artists like D. Eigler and D. Brodbent in their famous rendering of a nanoscale corral. Other established philosophers of science contest the images in nanoscience, arguing that renderings of the nanoscale are not "images" at all, indeed the very notion of "image" is irrelevant. See J. Pitt (2005) When is an image not an image, *Techné: Research in Philosophy and Technology*, 11(1): It is our contention that in NSR, images only contribute to physical intelligibility when seen as part of a triangle consisting of text, numerical information, and image.

[15] C. Toumey (2007) Cubism at the nanoscale, *Nature Nanotechnology*, 2: 587–589; C. Toumey (2009) Truth and beauty at the nanoscale, *Leonardo*, 42(2): 151–155; C. Toumey (2010) Images and icons, *Nature Nanotechnology*, 5: 3–4; M. Ruivenkamp, A. Rip (2010) Visualizing the invisible nanoscale. Study of visualization practices in nanotechnology community of practice.

[16] M. Sturken, L. Cartwright (2001) *Practices of Looking; An introduction to visual culture*. New York, NY: Oxford University Press.

[17] M. Ruivenkamp (2011) Circulating Images of Nanotechnology.

[18] D. Marr (1976) Early processing of visual information, *Philosophical Transactions of the Royal Society of London. B, Biological Sciences*, 275(942): 483–519; D. Marr, H.K. Nishihara (1978) Representation and recognition of the spatial organization of three-dimensional shapes, *Proceedings of the Royal Society of London. Series B. Biological Sciences*, 200(1140): 269–294; R.N. Shepard (1978) The mental image, *American Psychologist*, 33(2): 125; M. Denis, C.T. Greenbaum (1991) *Image and Cognition*. New York, NY: Harvester Wheatsheaf.

similarities between image portrayal and ontology are all that one can reasonably claim.[19] In this book, we discuss images in NSR in terms of information-communicating vehicles. To the extent that they deliver reliable information on form and properties, their resemblance with physical objects is of secondary importance. Images promote processes of knowing.

As indicated in our above reflection on form in NSR, the subjects of laboratory research are connected to the form of objects and to relations between forms. Here, images provide information about four aspects of nano-objects: (1) a view of a single object as a whole, (2) a view of the component parts of that object, (3) a view of the relationship between an object and surrounding like or different objects, and (4) a view of the object and what is interpreted as its environment. Note, each of these items is expressed as form. In nano, image and form comprise a unit that embraces observation, information, representation and, finally, understanding. It is here often impossible to disentangle the couple image/form. Today, some images of molecular and atomic entities are able to register dynamic states of changes in the form or position of objects. Seeing is thus multi-level and entails some visual qualities and impressions akin to seeing with the naked eye, where one can perceive an object and details and neighboring objects.

The detection and depiction operation of many of today's metrological and computational instruments that explore the nano world represent entities and clusters of entities in terms of a kind of architecture—a kind of landscape. Based on a repertory of points, lines, and curves, instruments generate three-dimensional images.[20] These are interpreted as representing atoms and molecules. Three-dimensionality is accompanied by fine-grain details and shadow, which confer a sense of palpable presence.

But precisely why do many NSR scientists choose images for purposes of data detection and organization, representation, analysis, and communication in the course of their research work? There certainly exists a plethora of alternatives. Since the early 1990s, thousands of researchers, who earlier in their career had reasoned and communicated in the framework of numerical values, curves, graphs, diagrams, or sketches, etc., have now switched to image-based information and representations. What is it in an image that is judged by nanoscientists as offering additional value over other categories of expression, representation, and reflection? What are the epistemological implications of images as a cognitive resource and platform?

[19] O. Bueno (2006) Representation at the nanoscale, *Philosophy of Science*, 76: 617–628; O. Bueno (2011) When physics and biology meet: The nanoscale case, *Studies in History and Philosophy of Science*, Part C, 42(2): 180–189; O. Bueno (2008a) Scientific representation. Microscopes and mathematical models. In: C. Bas van Fraassen (ed.) *Scientific Representation: Paradoxes and perspective*. Oxford: Clarendon Press.

[20] These images are not to be understood in terms of "likeness" and likeness criteria as recommended by van Fraassen (Bas C. van Fraassen, 2008) *Scientific Representation*). We adhere to Bueno's understanding of visual representation in NSR, where he stresses the link between representation and cognitively and technically useful information (O. Bueno, 2006) Representation at the Nanoscale; O. Bueno (2008) Scientific Representation, Microscopes, and Mathematical; O. Bueno (2008b) Visual evidence at the nanoscale), *Spontaneous Generations: A journal for the history and philosophy of science*, 2(1): 132.

An ensemble of seven considerations intervene: time, data collation, psychology and intuition, prospecting, analysis, formatting of findings, and presentation.[21] One nanophysicist complained that pre-image research absorbed huge quantities of time. Work grounded on images as data, representation, and output is far less time consuming, allowing rapid progress. Data in the form of numerical values prove slow and difficult to organize properly. Relations between streams or sets of data must be determined, and this alone may demand more mathematical operations. Much of this is circumvented with images which offer arrangement of numerical information that in principle collate with aspects of the objects under investigation.[22]

In pre-image efforts, it often proved quite difficult to generate a mental representation of the physical entity under study—to prospect it, grounded on the patterned data. This is not the case with images, which are seen as immediately offering access to entities.[23] Through an image, we were told, practitioners enjoy an intuition into their objects. In some instances, contact is so proximal that one almost has the feeling of touching it. There sometimes exists a strong presence of the "real." The image engenders a kind of psychological shock. As indicated earlier in this book, how many times were we told of the shock experienced in first "seeing" the image of the letters "IBM" spelled out by the spatial arrangement of single atoms?

Not least of all, images of objects, claim nanoscientists, engender a multi-level window onto entities. They can be observed in tiny detail—as segments. They can also be viewed at distance—as a synthetic whole. This twin vantage point permits a variety of visual perspectives that translate into cognitive perspectives.

In pre-image practice, decisions had to be taken about how best to structure information in order to communicate it. The choices were graphs, values, etc. Image-grounded results reduce the exigencies of communication. First, the question is not between image versus values. Most images are accompanied by listings of data that are automatically produced. Moreover, the intelligibility of an image for people working in the same field and possessing a minimum of information is often immediate, requiring little or no elucidation.[24] Pre-image representations were often a time of discomfort. How should the

[21] D.P. McCabe, A.D. Castel (2008) Seeing is believing: The effect of brain images on judgments of scientific reasoning, *Cognition*, 107(1): 343–352.

[22] Interview of Gérald Dujardin (IPM, University of Orsay) by authors, 17 June 2008, 10 October 2008, 11 April 2010.

[23] The fit of data to image and of image to real entities remains problematic on innumerable grounds (M. Lynch 1985) *Art and Artifact in Laboratory Science: A study of shop work and shop talk in a research laboratory*. London: Routledge & Kegan Paul; M. Lynch, S.Y. Edgerton (1988) Aesthetic and digital image processing representational craft in contemporary astronomy. In: G. Fyfe, J. Law (eds.) *Picturing Power; Visual depictions and social relations*. London: Routledge, pp. 184–220; M. Lynch (1990) The externalized retina: Selection and mathematization in the visual documentation of objects in the life sciences. In: M. Lynch, S. Woolgar (eds.) *Representation in Scientific Practice*. Cambridge, MA, London; The MIT Press, pp. 153–186; M. Lynch, S. Woolgar (eds.) (1990). *Representation in Scientific Practice*. Cambridge, MA: The MIT Press; M. Ruivenkamp (2011) Circulating Images of Nanotechnology.

[24] Interview of Tristan Cren (INSP, University of Paris) by authors, 18 September 2007, 2 January 2008, 11 October 2008, and June 2010.

entity be indicated? What aspects should be deleted or stressed, and how? Is the representation adequate and lucid? Should an alternative draftsman be contacted for an alternative rendering? How much additional time would be swallowed?

Finally, the move to images is also linked with professional and institutional considerations. In many fields of science, and most particularly in NSR, it is almost an expectation, a norm, to introduce images into articles. Indeed, in some prestigious journals, such as *Nano Letters* and *Nature Nano*, images are stand-alone, an abstract preceding an article, or they accompany the abstract. Many individual scientists insert images into their homepage, and many nanolaboratories document and publicize their work using colorful and striking images.[25]

The preceding discussion refers not solely to metrology-based images, but also to simulation-derived images. The simulation results born of computational instruments also operate in a framework of knowledge through seeing. Simulation findings can take the form technically of numerical data expressed in graphs and tables, but they are often rendered as images. Why? Firstly, simulation images allow researchers to see the entirety of a set of complex findings, to separate segments of the results, and to see connections between the whole and the parts. As in the case of metrology images, in simulation, such complex observations and through this intelligibility are simply not so readily available with numbers. Scientists can ascertain detail and the whole. Secondly, seeing in simulation enables one to identify the presence of process and to describe and analyze it: a series of images can depict the dynamics of an evolution in a system, and allow the observer to understand at a single glance in which direction the process flows and how it takes place. In that sense, one can say that seeing and understanding a process shown in simulation images open the way to the perception of a dynamics in a physical system with a sense of its complexity. In seeing through simulations, there is the possibility of extending observation in the direction of prediction. This is due to the fact that one may observe the existence of a sequence—of some regularity in the structure and of the events that occur in it.

A set of simulation images arising from nanoscale computational research on the dynamical behavior of a microscopic fluid, carried out by Uzi Landman in 2000, to see whether the Navier–Stokes theory applied to nanoscopic systems, illustrates the power of imagery. The mathematical principles that describe fluid flow were set forth more than 150 years ago and are known as the Navier–Stokes equations. How do these macroscopic rules work when fluids and flows are behaving at the nanoscale? Do the same rules apply or, given that the behavior of materials in this size regime often has little to do with their macro-sized cousins, are there new rules to be discovered? Small systems are influenced by randomness and noise much more than larger systems. Uzi Landman[26] reasoned that modifying the Navier–Stokes equations to include stochastic elements—this gives the probability that an event will occur—would allow them to accurately describe

[25] For example: <http://www.insp.jussieu.fr/>; <http://www.med.unc.edu/biochem/kuhlman>.

[26] Interview of Uzi Landman (Georgia Tech University) by T. Shinn, 29 January 2008. Landman is Director of the Center for Computational Materials Science, Regents' and Institute Professor, and Callaway Chair of Physics at the Georgia Institute of Technology.

the behavior of liquids in the nanoscale regime. Simulation experiments show that the stochastic Navier–Stokes formulation does work for fluid nanojets and nanobridges in a vacuum. In this study, scientists simulated a liquid propane bridge, which is a slender fluid structure connecting two larger bodies of liquid, much like a liquid channel connecting two rain puddles. The bridge was 6 nm in diameter and 24 nm long. The goal was to study how the bridge collapses. The simulation focused on a model with a variable-gas-pressure environment surrounding the bridge. Under high pressure, the bridge tends to create a long thread and break asymmetrically on one side or the other of the thread, instead of in the middle. Simulation images show that the asymmetric breakup of the nanobridge in a gaseous environment relates to molecular evaporation and condensation processes, and to their dependence on the curvature of the shape profile of the nanobridge. A change in gas pressure modifies molecular evaporation "If the bridge is in a vacuum, molecules evaporating from the bridge are sucked away and do not come back," and condensation behavior, which affects the evolving structure of the bridge, said Landman. "But if there are gas molecules surrounding the bridge, some of the molecules that evaporate will collide with the gas, and due to these collisions the scattered molecules may change direction and come back to the nanobridge and condense on it."[27] As they return they may fill in spaces where other atoms have evaporated. In other words, the evaporation–condensation processes serve to redistribute the liquid propane along the nanobridge, resulting in an asymmetric shape of the breakage. The higher is the pressure surrounding the bridge, the higher the probability that the evaporating atoms will collide with the gas and condense on the nanobridge.

The four images shown in Figure 4.1 depict the evolution of the form of a nanostructured fluid (indicated by dark gray in images (a) and (b), and by a lighter gray in images (c) and (d)) under evolving conditions of pressure, as described above. Image (a) represents the liquid, consisting of molecules in dark gray, in an environment where the gas (white atoms) pressure is low. The morphology of the band of liquid consists of a regular extended rectangle. In image (b), the gas pressure is higher, the edges of the liquid band are irregular, and the band is thinner. The mix of gray and white spheres indicates an evaporation/condensation circular dynamic, to a slight degree. Image (c) shows that the bridge is becoming narrower and more irregular. In image (d), the liquid bridge is broken, but because the liquid atoms condense, as a result of their collisions with gas atoms, they again adhere to the liquid bridge, and give rise to an asymmetric geometry.[28]

Simulation images of nanoscale events here give an intelligibility to processes at the nanoscale. They make each step within the process comprehensible through rendering of the form, force, and perturbation that constitute the dynamics of the object. Simulation images are highly detailed and systematic. It is worth noting that seeing and understanding through simulation frequently privileges a sense of regularity, organization,

[27] <http://www.nano.gatech.edu/news/release.php?id=1282>.

[28] <http://www.gatech.edu/upload/pr/tcp39571.jpg> (Fluid Dynamics Works on Nanoscale in Real World) <http://www.nano.gatech.edu/news/release.php?id=1282> (June–July 2012).

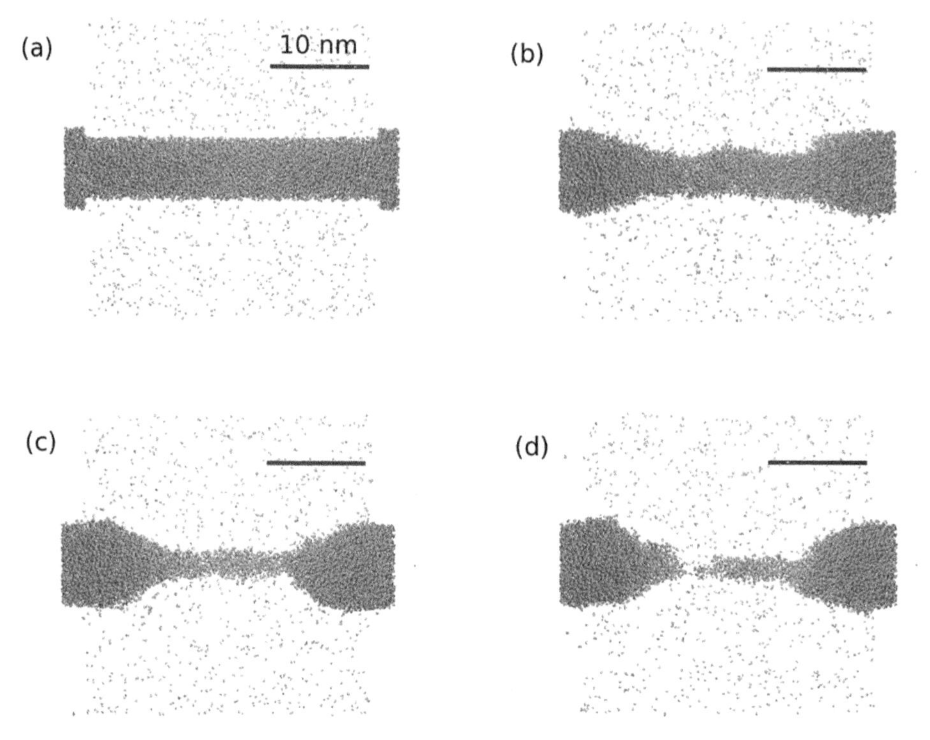

Figure 4.1 Fluid dynamics works on the nanoscale in the real world. Image reproduced with permission of Uzi Landman, Georgia Institute of Technology.

and structure in three dimensions. Such three-dimensionality can be important to NSR because of the very spatial properties (even spatial determinism) of nano-objects and their forms.

Images, be they computational or metrological, are considered by most researchers who were interviewed for this study to constitute a reasonable guarantee that what they depict warrants discussion and constitutes an acceptable basis for reasonable attempts at understanding. This is certainly not to say, though, that difficulties do not sometimes intervene in image production or manipulation, which fitfully proves that they remain problematic under certain circumstances. The validation of any novel detection and observation technical system, such as scanning probe microscopy and nanosimulation, is a process requiring vast experience and time. Nano-imagery is nevertheless seen by most scientists to provide a sufficiently robust and exacting basis for the description of their research-object.[29]

[29] Interview of Vincent Dubost (INSP, University of Paris) 30 April 2008; interview of Tristan Cren (INSP, University of Paris) 5 May 2008; interviews of Jacques Jupille (INSP, Paris) 17 September 2008, Catherine Gourdon (INSP, University of Paris) 20 October 2008, and Gérald Dujardin (IPM, University of Orsay) 14 April 2009, by authors.

The computational images that are produced to depict processes occurring at the nanoscale show that, in the case of the dynamics of fluids, the Navier–Stokes laws do not apply in the same way that they do on the macro-scale. In that sense, an epistemology that systematically addresses the issues of dynamics emerges.

Scientific research in nanoscience gives rise to three categories of images, all of which are computer dependent: primary images, secondary images, and computational simulation images.[30] Primary images are produced by measurement instruments which acquire data that are then transduced by a specialized algorithm linked to a computer which in turn generates a topological or associated depiction of the object under investigation.[31] The scanning tunneling microscope and the atomic force microscope are typical devices that produce primary images.[32] Instruments and their image packages are purchased equipped with their own imaging algorithms. While some variability in algorithms is possible, most scientists retain the initial package.[33] The technical details relating to the production of these images entail relatively few decisions about settings. The spatial information offered by the instrument remains unmodified. It is judged by its users to be the most adequate and valid representation of the phenomenon under examination. Here, minimally intervened metrology of topology and related parameters are the foundational information unit of image production and cognition. There is thus a kind of standardization in the possibilities and expectations of the architecture and constraints of primary images.

The secondary images of NSR, our second class of imaging, issue from the primary images and systematically retain their foundational data.[34] They require the introduction of a computer graphics program specialized in image processing.[35] Many such

[30] C. Allamel-Raffin (2004) La Production et les fonctions des images en physique des matériaux et en astrophysique. Doctoral dissertation, University of Strasbourg; C. Allamel-Raffin (2006) La Complexité des images scientifiques. Ce que la sémiotique de l'image nous apprend sur l'objectivité scientifique, *Communication et langages*, 149(1): 97–111.

[31] It would be misleading to consider that detection and the modeling associated with primary images are unproblematic. See A. Mohebi, P. Fieguth (2006) Posterior sampling of scientific images. In: A. Campilho, M.S. Kamel (eds.) *Image Analysis and Recognition*. Berlin, Heidelberg: Springer, pp.339–350; J. Hennig (2005) Changes in the design of scanning tunneling microscopic images from 1980 to 1990, *Techné: Research in Philosophy and Technology*, 8(2).

[32] C. Mody (2006) Corporations, universities, and instrumental communities: Commercializing probe microscopy, 1981–1996, *Technology and Culture*, 47(1): 56–80; C. Mody, M. Lynch (2010) Test objects and other epistemic things: A history of a nanoscale object, *British Journal for the History of Science*, 43(3): 423–458; C. Mody (2011) *Instrumental Community: Probe microscopy and the path to nanotechnology*. Cambridge, MA: The MIT Press; S. Loeve (2009) Le Concept de technologie à l'échelle des molécules-machines. Philosophie des techniques à l'usage des citoyens du nanomonde. Thèse de Doctorat de philosophie, d'épistémologie et d'histoire des sciences et des techniques, Université de Paris-Ouest.; M. Ruivenkamp, A. Rip (2010) Visualizing the invisible nanoscale study, *Science Studies*, 23(1): 3–36.

[33] Y.G. Lee, K.W. Lyons, S.C. Feng (2004) Software architecture for a virtual environment for nano scale assembly (VENSA), *Journal of Research of the National Institute of Standards and Technology*, 109(2): 279–290; M. Sitti, S. Origuchi, H. Ashimoto (1998) Nano tele-manipulation using virtual reality interface. *IEEE Xplore*, 1(1): 171–176; C. Robinson (2004) Images in nanoscience/technology. In: D. Baird, A. Nordmann, J. Schummer (eds.) *Discovering the Nanoscale*. Amsterdam: IOS Press, pp. 165–169.

[34] C. Allamel-Raffin (2004) La production et les fonctions des images en physique des matériaux.

[35] J. Lawrence, T. Funkhouser (2004) A painting interface for interactive surface deformations, *Graphical models*, 66(6): 418–438.

programs are available commercially, for example, PaintShop. PaintShop is emblematic of the computational information-observation revolution of the 1970s, 1980s, and 1990s (see Chapter 3). These programs are typically run on conventional desktop computers of the sort sitting on almost every laboratory desk. Our nanoresearcher interviewees tell us that processing linked to secondary images is performed for observational purposes.[36] The optical information of primary images is such that it is sometimes difficult to disentangle. Data may be tightly packed and intertwined. This appears to be particularly the case for the morphology-rich NSR study of molecular surface features and multi-molecule spatial dispositions. Reprocessing of information allows the introduction of visual effects: an item that optically is barely perceptible in the background may be moved to the foreground. A target item lost in a jumbled cluster can be isolated. An object can be magnified in order to see certain features. The existing relief of an object can be amplified to make it clearer and more readily explored.

Finally, color is part of the arsenal of NSR secondary images.[37] It is employed in order to differentiate the various parts of an image, and to distinguish the phenomenon under study from its environment. Beyond these manipulations for narrowly cognitive ends, colors are also introduced for the purpose of scientific communication, to make images more appealing to a broader public and for aesthetic effects.

The work engaged in producing secondary images is often quite time consuming. It consists of much tinkering, as scientists introduce one computer-graphics command after another, in an attempt to obtain the wanted image clarification.[38] If one command fails, another is attempted. In many instances the result does not improve the view of the target, so an alternative procedure must be attempted. One can speak here of a type of "tatonnement" ("feeling one's way"). To repeat: information from the primary image is not changed, only the capacity to enhance visual information contained in the initial rendering is affected with these secondary images. Notwithstanding, in nanoresearch, criticism can arise among steadfast practitioners of primary images that reprocessing constitutes a kind of computer game that threatens the authenticity of aspects of physical intelligibility.[39]

[36] L. Wojnar (1999) *Image Analysis; Applications in materials engineering*. London: CRC Press.

[37] M. Farge (1990) L'Imagerie scientifique. Choix des palettes de couleur pour la visualisation des champs scalaires bidimensionnels, *L'Aéronautique et l'astronautique*, 1(140): 24–33; J.W. Goethe (1980) *Le Traité des couleurs*. Paris: Éditions Triades; I. Hacking (2006) An other world is being constructed right now: The ultra-cold. Conference "The Shape of Experiment Berlin," 2–5 June, Max Planck Institute for the History of Science; T. Welsh, M. Ashikhmin, K. Mueller (2002) Transferring color to greyscale images, *Proceedings of the 29th annual conference on Computer graphics and interactive techniques ACM Transactions on Graphics (TOG)–Proceedings of ACM SIGGRAPH*, 21(3): 277–280.

[38] Interview of Vincent Dubost (INSP, University of Paris) by authors, 8 March 2008.

[39] A. Mohebi, P. Fieguth (2006) Posterior sampling of scientific images. Workshop "Nano–objets synthétiques et Bio–inspirés," Université Orsay Paris-Sud. 20–21 January 2011; Interview of D. Pompon and A. Laisné (Centre de Génétique Moléculaire). LIMSI (Gif/Yvette) by T. Shinn and Anne Marcovich, 15 March 2011.

Finally, the third category of images prevalent in NSR, what we refer to as computational simulation images, represents computational output as form.[40] Computation thus operates on two levels: it calculates physical phenomena and then, in a second phase, that output is numerically processed through algorithms and emerges as images. This computational imagery in science was first developed in the 1960s, often in connection with molecular graphics.[41] The movement grew in importance during the 1980s and today it is widespread throughout nanoscale research. In recent years, simulation results in science have almost always been expressed as images. They are particularly abundant in NSR. Five features stand out: (1) There is a kind of standardization that traverses almost all simulation images due to the generative algorithms. For example, a lattice is always structured in the same way; proteins are represented as ribbons. (2) The scale of the object being represented is specified in terms of dimension and distance. This permits particular appreciation of spatial relations. (3) Simulation images offer a high level of detail. For a given phenomenon, the number of atoms can be counted and the element is specified. One can identify a defect for example. (4) The different elements that appear in a simulation image are presented in a crisp and distinct fashion. Entities are always neatly presented and well distinguished. (5) The components of simulation images are not just exact, they are also exacting in presentation. It almost appears that they have been cleansed and purified in order to enhance readability and to forge understanding. They are nevertheless not to be mistaken for idealizations! The clean aspect of simulation-experimentation images is sometimes a consequence of the choice of the variables introduced in an upstream analytic program. It is not a product of the imaging program. Because of the relative high speed and low cost of simulation based experiments, it is possible to generate a large number of images where each image represents a modification in the empirical data introduced to the program. It is easier to do a simulation that requires days, weeks, or months, than it is to carry out some metrological experiments that require years of preparation. Unlike the above-mentioned secondary images, scientists do not reprocess computational simulation images. Modification of images derives exclusively and directly from the introduction of new values or parameters. In many instances, what is judged by both metrological experimenters and simulation specialists to be a successful simulation image is one that corresponds with a metrological image generated in the course of a metrology physical experiment. Simulation practitioners working with metrology experimenters frequently request that the latter first provide an image as a starting point for the choice of programs and selection of empirical values. This is intended to guide

[40] F. Rohrich (1990) Computer simulation in the physical sciences, *Proceedings of the Biennial Meeting of the Philosophy of Science Association*, 2: 507–518; J. Lenhard (2004) Nanoscience and the janus-faced character of simulations. In: D. Baird, A. Nordmann, J. Schummer (eds.) (2004) *Discovering the nanoscale*, pp. 93–100; G. Küppers, J. Lenhard (2006) Simulation and a revolution in modelling style: from hierarchical to network-like integration. In: G. Küppers, J. Lenhard, T. Shinn (eds.) *Simulation: Pragmatic constructions of reality*. Dordrecht: Springer, pp.89–106.

[41] E. Francoeur (2004) From model kits to interactive computer graphics. In S. de Chadarevian, N. Hopwood (eds.) *Models. The third dimension of science*. Palo Alto, CA: Stanford University Press, pp. 402–433.

the efforts of simulation. The range of analytic scenarios available to simulation is so considerable, that it is advisable to first anticipate in which direction to focus.

In our view of epistemology, a kind of epistemology that the philosopher of science, Ronald Giere, refers to as "naturalistic epistemology," the entire sweep of endeavors associated with cognition generation is included.[42] This notion is expressed in our combinatorial concept of epistemology/image. Imaging is central to metrological and computational experimental work. It provides the bridge between data and perception. It offers a key category, form, necessary for processes of analysis. Images are the vocabulary of discussion and debate, and they constitute an important medium of communication between scientists in NSR and broader modes of communication.

4.2.2 How images are worked out

In the case study that follows, we explore how images operate in one area of nanoscience, how the combinatorial of form and image develops, and what this means for the epistemology of nanoscale research. Below is a detailed report of a combined STM and STS (scanning tunneling spectroscopy) exploration of a nanoscale superconductor in a strong vortex confinement regime where images and form are central. While much has been learned about superconductivity since the discovery of the phenomenon by Heike Kamerlingh Onnes in 1911,[43] study continues on this family of events, and particularly on the dynamics and environment of superconducting vortices at the nanoscale.

Here we will examine scientific images in connection with form-driven knowledge related to this topic. In this low-temperature study, the object of investigation is itself "created," and the environment does not appear to constitute a factor that would corrupt the study.[44] Eventual "experimental artifacts" would not derive from environment, but instead from the experimental set-up proper. Our discussion of the laboratory images first deals with images associated with the opening stages of investigation.

The aim of this research was to identify the form, position, and movement of the vortices, their internal forms, and to try to understand the physical forces that underlie the whole. The experimental apparatus was built and tested over a period of about five years by the participating team, directed by physicist, Dimitri Roditchev, at the Institut des Nanosciences de Paris (INSP). It involved a novel STM tip, improved camera, low-vibration pumps and plumbing, and a long-duration cryostat. The advantage of this instrument over competitors is that it is highly stable, generating little noise; it allows

[42] R. Giere (2010) *Explaining Science. A cognitive approach*. Chicago, IL: University of Chicago Press.

[43] J. Matricon, G. Waysand (2003) *The Cold Wars: A history of superconductivity*. New Brunswick, NJ: Rutgers University Press; K. Gavroglou, Y. Goudaroulis (1989) *Methodological Aspects of the Development of Low Temperature Physics, 1881–1956: Concepts out of context(s)*. Dordrecht: Kluwer Academic Publishers.

[44] M. Lynch (2006) The production of scientific images. Visions and re-vision in the history, philosophy and sociology of science. In L. Pauwels (ed.) *Visual Cultures of Science: Rethinking representational practices in knowledge building and science communication*. Dartmouth, NH: Dartmouth College Press, pp. 26–41; M. Lynch (2006b) Discipline and the material form of images. An analysis of scientific visibility. In: L. Pauwels (ed.) *Visual Cultures of Science*, pp.195–232.

experimental runs of long duration; it offers advantages in experimental control. This research program privileges the use of images as a method of information acquisition, and particularly their employment in the analysis of phenomena. Several members of the team are highly experienced in the production of secondary images of nanoscale objects, the interpretation of these images, and their reflection on the findings incorporates a kind of image-driven epistemology. The passage from primary image to secondary image in the work of the researchers is neither distinct nor readily discernible. Features that are frequently associated with primary images appear to be transgressed, as processing may occur at a very early stage of image production.

The study of a nanovortex in superconductivity, portrayed here, entails twin sources of information: STM measurements give the topology of the object; scanning tunneling spectroscopy registers the intensity of the relationship of the current with the voltage.[45] These two devices give rise to two images and a third image combines images 1 and 2.[46]

The initial image (Figure 4.2) depicts a nanoscale superconducting region that appears as an island in a non-conducting sea; it shows a trough in the nano-island that constitutes the superconducting vortex. The scientists' study focuses on the position, dimension, and dynamics of this vortex. In order to more closely scrutinize the vortex trough, in this image they re-proportioned the optical-relative dimensions of the island and depth of the trough by accentuating the thickness of the former. The researchers modified optical impressions in order to more fully render the reality of the phenomenon. This image consists of a detection grid of 200 points by 200 points. Each individual point indicates the height of the phenomenon measured by the STM's tip. The resulting image is

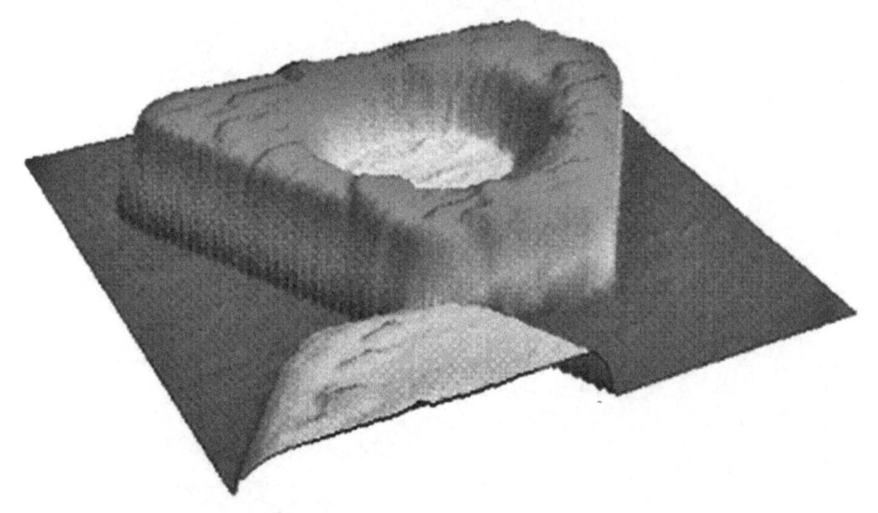

Figure 4.2 STM image of the nanovortex. Image reproduced with permission of Tristan Cren (INSP).

[45] T. Cren et al. (2009) Ultimate vortex confinement studied by scanning tunnelling spectroscopy, *Physical Review Letters*, 102, (1), pages: 74–78.

[46] These images have been kindly transmitted by Tristan Cren (INSP, University of Paris) to the authors.

three-dimensional with artificial colors added. For editorial reasons the scientists' images that in their published text appeared in color for intellectual and communicational purposes, are here depicted in black, gray, and white. In the scientific publication height is indicated by colors, but is here shown by different shades of gray.

In addition to this topological data-collection, the scientists generated images based on spectral data. For each of the data points of the preceding, a tunneling spectrum was recorded, i.e. the tunneling current "I" was measured as a function of the tunneling voltage "V." This gives a spectrum I(V) (320 different values of this voltage were measured), and it gives a measure of the force of the magnetic field. These measurements allow identification of "normal"—as opposed to regions that are characterized by superconductivity.

This gives rise to the following image (Figure 4.3). This spectroscopic-based information is depicted in two dimensions, as opposed to the three dimensions of the image in Figure 4.2. The information in the two-dimensional rendering cannot visually signal three dimensions and thus cannot indicate the trough optically. It nevertheless reveals the position of the trough on the island, through the introduction of color to signal its form. The violet of the trough and of the sea (indicated in our depiction by the regions b) contrasts with the purple of the island (in our image indicated by a). Color thus functions to

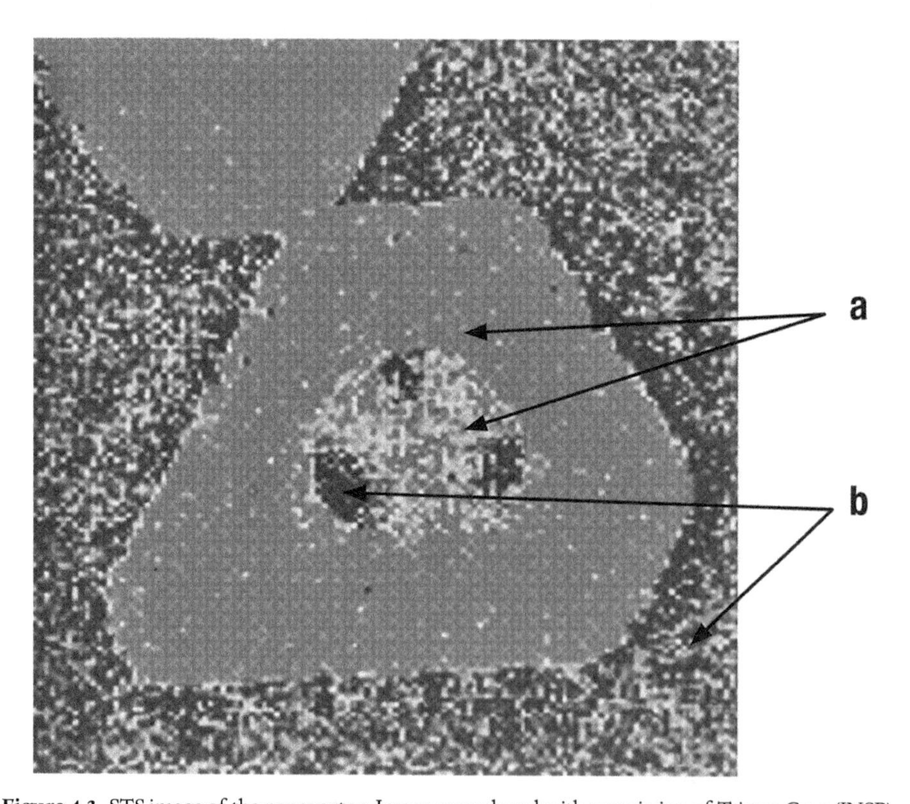

Figure 4.3 STS image of the nanovortex. Image reproduced with permission of Tristan Cren (INSP).

distinguish the different spaces; it also indicates the superconductivity state of each area (in the published article, red for superconductivity, indicated here by the letter a), violet for the non-conducting state (normal state) here signaled by the letter b) (Figure 4.3). This image thus above all reveals the patchy heterogeneity of superconducting regions.

The final operation (see Figure 4.4) combines the three-dimensional topographical image of the STM image (Figure 4.2) with the two-dimensional spectroscopic image (Figure 4.3) that represents the forces of the magnetic field. The distribution of colors as markers of a specific node of nanoscale superconductivity (Figure 4.3) are superimposed on the initial STM image (in Figure 4.2), whose three-dimensionality provides information about the geometry and depth of the trough—the object of investigation. The colors of the image in Figure 4.3 were smoothed in order to attenuate the tiny variations of voltage in the non-conducting regions surrounding the superconducting nano-islands. This constituted troubling background noise. Why was this done? The environment around the islands added no relevant information and risked distracting the eye of the viewer.

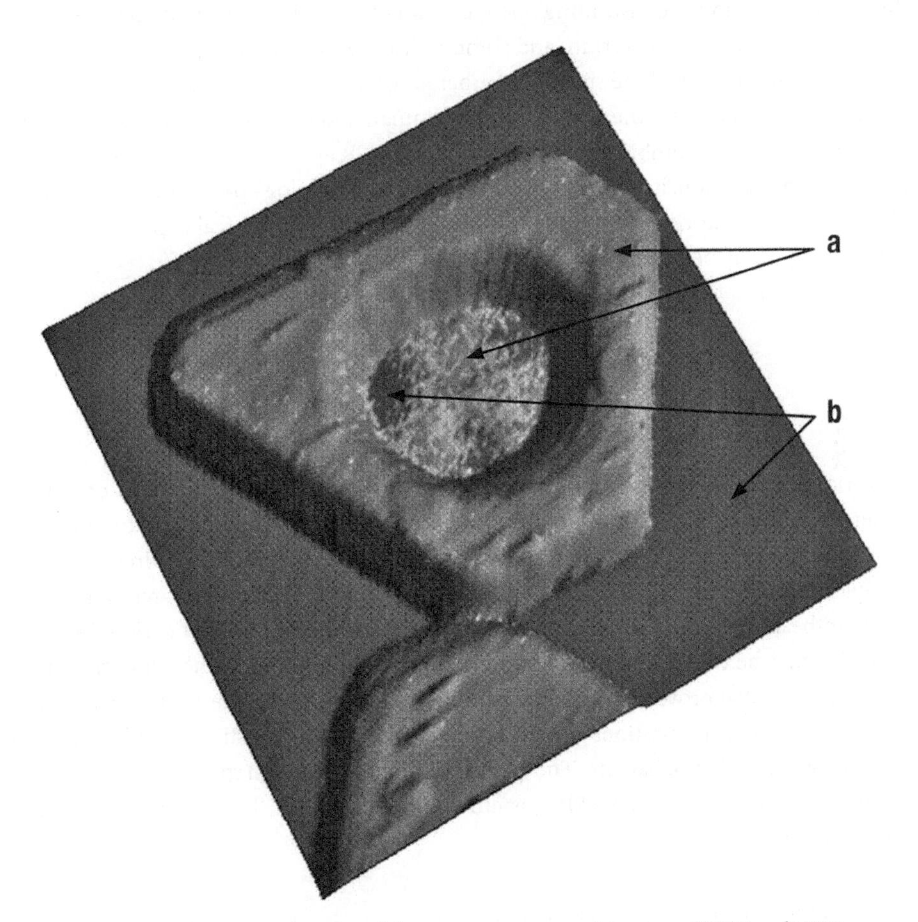

Figure 4.4 Three-dimensional image of the nanovortex: a combination of STM and STS images. Image reproduced with permission of Tristan Cren (INSP).

The 3D image is also modified; the perspective is slightly shifted in order to better see inside the island where items of interest occur. In effect, visual change was introduced, but it did not alter the information collected during the experiment. The resulting image (Figure 4.4) dramatically exhibits the correlation between topography and spectroscopy. "All is shown in the image: no need for a discourse. The image speaks for itself for those who minimally know the subject."[47]

Commentary such as this characterizes the relatively unproblematized stance of scientists on closure of a research cycle. The image is regarded as independent of the many decisions taken during its production and processing. The image is not simply autonomous of its conditions of production, it is viewed as depicting the elements under investigation as self-evident within the frame of the image.

In this last quotation, proffered by our informant, one confronts the difficult problem in the study of scientific images of trying to un-entangle image as information, as a vehicle for analysis and interpretation, and image as a medium of communication.[48] One can think of information as constituting one spectrum of analysis, interpretation as a second but loosely and indirect spectrum, and communication as constituting a third spectrum. It is form that holds the three spectra together and gives them meaning. The three spectra are a pre-condition for the intelligibility of human understanding, and in this way the image is an explicitly combinatorial cornerstone of NSR epistemology. This last observation raises the question: is one key attribute of NSR epistemology perhaps the crucial role played by combinatorial elements? More than other epistemological expressions, NSR epistemology may be "a combinatorial epistemology."

4.3 THE QUESTION–MATERIAL TO MATERIAL–QUESTION SPIRAL

Historically, in most scientific research the confines engraved in the materials of nature generally prescribe the questions formulated by researchers—from matter to question. In NSR, however, this flow between material and question becomes multi-directional and complex: scientists now freely raise questions and putatively relevant research-objects are then synthesized in order to constitute the appropriate terrain of investigation. NSR entails what may be viewed as a novel configuration of research cycle, where questions produce materials that provide answers, which in turn generate new questions. This circular relationship between questions–material, and then material–questions, yields discovery of new structures and behaviors. The question–material to design spiral represents a constitutive combinatorial of the NSR epistemological matrix which frequently fuels entire chapters of research.

[47] Letter from Tristan Cren to authors (22 October 2010).

[48] M. Lynch (2006) Discipline and the material form of images; M. Ruivenkamp, A. Rip (2010) Visualizing the invisible nanoscale study.

Why is the question–material system epistemological in character? Firstly, as a combinatorial, it constitutes a novel field of reflection and practice. Secondly, in the absence of this combinatorial, progress is difficult and even improbable. Thirdly, the pair question–material constitutes an entity that entails its own internal logic and which must necessarily be examined and used accordingly.

The dialogs question–material and material–question generate two additional related configurations. In "from material to design," nanopractitioners employ an epistemology based on intended outputs where control is central . Outputs are not, however, intended as products, but instead as a cognitive validation of the interface between design research and the potentialities of a given material. Finally, in the "materiality of matter," there emerges a variety of epistemology that stresses the operation of aspects of nanostructured matter that are here judged essential, yet do not routinely figure in much orthodox nanoresearch and reflection. This epistemology of the materiality of matter may be indirectly seen as evocative of issues of meta-causality.

4.3.1 From matter to question

New materials provoke novel questions. The recent emergence of new nanostructured semiconductor magnetic materials has opened the way to a wave of new questions. Emblematic of this is the discovery of giant magnetic resonance, pioneered by Albert Fert, who received the 2007 Nobel Prize in physics. Many nanophysicists are now formulating additional questions based on recently established magnetic semiconductor substances. The rise of thin-layers technology in semiconductors has permitted the consideration of new questions about magnetic resonance fields and about electron–spin behavior. Using a novel semiconductor substance, gallium manganese arsenic/gallium indium arsenic, a nanophysicist, Catherine Gourdon,[49] working at the INSP, has enriched this experimental material through the introduction of crystalline islands and stumps. Her goal was to create a material adapted to the production of "magnetic domains," with the objective of studying domain position and dynamics. The term magnetic domain refers to the magnetization of a piece of a material, which spontaneously divides into multiple regions. The regions separating magnetic domains are called "domain walls." The novel substances and domain perspective permitted Gourdon to raise questions about the correlation between the rapidity of introduction of magnetic fields into the semiconductor material, and the organization of lamellar domains around the region where the magnetic field penetrates the material. She also studied the relations between multiple domains. She explored the speed of extension of domain walls in relation to the magnetic field. Mechanical characteristics of the semiconductor were evoked in order to study magnetic domains and their corresponding walls.

In this episode, using her novel substance, the NSR practitioner raised questions concerning the geography of domain walls, their dynamics in terms of speed of introduction,

[49] Interview of Catherine Gourdon (INSP) by authors (23 February 2008).

and the relation between materials and the phenomenon. It was the availability of a novel material that constituted the pre-condition for this questioning. A new material meant fresh formulations of a long-standing important research question. While she fell short of fully answering her range of interrogations, in the course of the project she nevertheless came to see that, in her particular field, the dynamics of domain mobility continued to resist research efforts. In summary, in this case study, the material raised far more, and more complex questions than could be managed.

4.3.2 From question to material

In the case of one of the scientists whom we repeatedly interviewed, Gerald Dujardin at the University of Orsay (already encountered in Section 2.1), the practitioner spoke at length of his investigation involving a dialog between question and material.[50] For him, questions about the architecture and motion of single molecules and the organically related issue of appropriate materials, constitute a defined world of research and knowledge. In this world, modifications of the structure and operation of substances, and even the invention of new ones, are all part of research, and it is very precisely in this way that transformation of materials penetrates epistemology.

In this example, the answer to the question of how single molecules are structured and how they behave under certain conditions entails control. Stated differently, formulation of a question, and addressing it properly, incorporate an understanding of the material per se, beyond the immediate question, and also revolve around processes of control. For Dujardin, although he opens a research project with a question, the interrogation necessitates consideration of materials and acquisition of control. Control, material, and question together form an inexorable dynamic chain.

Dujardin's initial question focused on the external and internal transformations induced in a single molecule, the forces entailed in transformation, and the location inside the molecule of the necessary transformational stimulation. This required a highly specific substance capable of morphological structural change under defined conditions. So the query meant cognition about form, movement, and the dynamics of action. The research associated with obtaining results had to be preceded by a long period during which the acquisition of an appropriate substance and an understanding of how to control it were foremost. In practical terms, this meant learning just how and where to excite a single molecule using a strategically situated STM tip.

> For an entire decade, we strove to manipulate single molecules. To manipulate the molecule, we stimulated it using the electrons emitted by the STM tip: we strove to make it move, to break it, to disassociate it, to make it recombine. For me it was a transposition of my earlier work on a large number of molecules—and here I was working with a single molecule.

[50] Interview of Gérald Dujardin (ISM, University of Orsay) by authors, 20 May 2008 and 15 September 2008.

From this quotation, one sees that architecting and description are frequently objectives in themselves in NSR. Through control, the goal is then to augment knowledge about a given material at the nanoscale.

> The electrons of the STM tip allow excitation of a certain site of the molecule. One can control a huge amount of things. We are now able to control reversible movements, displacements, translations, rotations, changes in the form of the molecule. At the beginning it was rudimentary: one bombed the material and broke the molecule. It was irreversible. Now, we can do much more precise things, one can control such and such movement rather than this other one. We can generate reversible movements and that's much more interesting.

The resulting capacity for control freed the researcher to contemplate his field from a broad perspective. Materials would no longer constitute a constraint to cognition. Dujardin was now free to translate his questions into experiments, and on the basis of experimentation to generate answers to his interrogations. The physicist could address the important questions that interested him in the confident knowledge that he could build the substances necessary to answer them.

What answers to his long-standing questions did Dujardin harvest? He learned that for some families of single molecules there exists a specific site around which segments of the substance swing in a reversible manner, and that the location of the dynamic components is predictable and constant. Motion of the mobile segments is only in part determined by the location of exogenous stimulation and by the forces that are introduced. A second factor also proves essential, namely the environment of the molecule. The presence of certain surrounding substances in a particular location powerfully affects the behavior of the molecule, to an extent that the dynamic movement is cancelled. One key reply to his initial query lay in the discovery that the selected molecule acts in a bipolar manner. It shifts regularly from one predictable position to another. It may be conceived as behaving like a kind of switch. Dujardin's questions were not guided by considerations of a technological function—such as a switch. Even near the close of his project when this perspective could be envisaged, it never became the horizon for Dujardin. However of utmost interest, the physicist's preoccupation had included questions of what he terms "functionality." For Dujardin, there exists a great distinction between his functionality and technological functions. "Functions" are connected to effects: in contrast, "functionality" is to be grasped as the possibility of certain kinds of action, along with an understanding of how those actions are physically produced, and in what sort of system. "Functionality" is hence part of cognition. The complicated relation between design and materials, and between functionality and function, along with their epistemological thrust and implications, will be discussed in the next section.

4.3.3 From material to design

In the case presented below, this orientation associates three activities: (1) selection of a material whose properties can enable experiments which promote the capacity to generate numerical operations (computations); (2) design and constitution of an appropriately functional architecture of the relevant material; (3) translation of the materialized form/structured objects into intelligible computational information. In this configuration, experimentation focuses on a nanomaterial and the organization of its constituents that enable an algorithm-based encoding and decoding procedure. We propose that this kind of endeavor reflects an indirect materials-to-design oriented epistemology. This is present in the research projects of Eric Winfree, who has employed DNA substances and architectures to create forms that carry computational algorithms capable of performing calculations.

Winfree was trained as a mathematician at Caltech. He is now a professor at this university, where he directs a multi-disciplinary laboratory. His appointment in this capacity constitutes the institutionalization and legitimizing of research on the employment of certain biosubstances for their computational algorithmic properties. Winfree won the Feynman Nanotechnology Prize in 2006, together with Paul Rothemund, for their theory of molecular computation and algorithmic self-assembly. In the words of Winfree, "One would never have thought of this thirty years ago; and such kinds of investigation clearly demonstrate the significance of design for the study of properties—and not least of all, at the nanoscale."[51]

Winfree designs the shape of DNA crystals for the development of information processing. He uses DNA structure, dynamics, and logic to generate algorithmic properties. Why select DNA material to achieve computational ends, and how does the DNA lead to computational capacity?

DNA was relevant for Winfree because it can self-assemble in a repetitive, selective, and combinatorial way. Repetitive self-assembling crystal DNA constructions are rigorously self-duplicating. They can be organized in two-dimensional motifs, comparable to "tiles" whose geometry may intersect as in a puzzle. This puzzle allows two possibilities: (1) the pieces fit, (2) the pieces do not fit. This corresponds to the zero/one computational system. These DNA combinations can be considered as gates, for the passage or impossibility of information. Winfree employs the DNA as a dynamical object whose form is expressed as a computational property, capable of systematic information organization.

Each tile measures two by four by 12 nm and can have either a square or a hexagonal shape. Form is essential in a second sphere: the edges of each tile possess an irregular conformation. These little bumps are the sites of specific bits of information.

In the words of Winfree:[52]

> If you think of your bathroom tiles, that works, except, let's say you have an interesting bathroom where the tiles aren't just square, they've got funny bumps on the edges. There might be several different kinds of them. If you think of a jigsaw puzzle, those are also tiles. In some jigsaw puzzles, every edge is slightly different. In other kinds of jigsaw puzzles, some of the edges are the same, and you have to figure out which piece goes where by some kind of more global way that things fit together. Geometrically, that's what a tile is. A molecular tile is just

[51] Interview of Erik Winfree (Caltech) by Terry Shinn, 30 January 2008.
[52] Interview of Erik Winfree (Caltech) by Terry Shinn, 30 January 2008.

> a phrase we use to talk... What is exciting to people is that it's such a flexible building-block for controlling that on a nanometer scale.

Winfree continually likens the computational properties of DNA to computer programs.

> You can control that by designing a program, which is a set of tiles that basically has that logic built into how they fit together. In some sense, it's analogous to how the lines of a program get executed, except in this case, just the shapes are the encoding of the program. They say, "If you see under this circumstance, put that kind of object here." That's sort of like a computer program that says, "If these conditions hold, put the data over there." It's the same thing, but in DNA crystal growth.

In the epistemology of "from materials to design," the architecting of a substance is foremost. A substance is chosen because its physical properties correspond with the embodiment of wanted operations. These operations are "functions" effects—that is to say intended concrete products—versus the above discussed "functionalities" expressed in the research of Dujardin (see Section 3.2). But while the intrinsic quality of the selected material is crucial, this is not sufficient. What counts is the design, the architecting of the substance. The function of the system revolves around the design. It is safe to say that, for all practical purposes, the selection of material is dictated by design requirements. Unlike the epistemological scenario of "from question to material," in "from material to design" there exists simultaneity of question and material. This notion of simultaneity is echoed in descriptivism and in what we refer to as "local" (still to be discussed). In other words, it privileges vistas characterized by overlap and by self-referencing proximity. A strong and constant locking of query and substance is here driven precisely by design criteria. This epistemology of NSR work is thus characterized by integrated interdependencies on multiple levels, where issues of sequence are strikingly absent. The possibility of such an epistemology is powerful in nano, because of the potential for materials synthesis and the capacity to pre-design forms and structure molecule by molecule and atom by atom.

4.3.4 Questions of materiality

The research of Yves Borensztein (INSP) on gold nanoparticle surfaces and their optical behavior raises a fascinating "essentialist" question rarely encountered in contemporary science: which causes underpin physical properties that are expressed through precisely defined characteristics in matter? The discussion revolves around "materiality." Reflection on materiality may be entertained because in nano, materials and their synthesis constitute a new territory in the history of science, where horizons remain open and where many unanswered queries persist. Among nanopractitioners, space for discussion and reflection over the issues of links between properties, cause, and the structure of substance have the status of crucial questions. Borensztein and his team's research figure is an instance of this essentialist current. It is in this context that Borensztein ponders questions of the link between matter structure at the nanoscale and its physical optical properties.

This linkage is what we refer to as "materiality," which inside the pale of nanostructured objects constitutes a curious yet captivating epistemological possibility.

As indicated throughout this book, the definition of nanoscience mainly revolves around size—today most often between, say, 1 and 30 nm. For example, the light-emitting properties of a quantum dot (its very existence and operation) is contingent on its nanodimensions, and this is equally true of the photon behavior of nanowells and nanowires. In his research work on the optical properties of nanofilms and nanoparticle surfaces, Borensztein insists that it is issues of form, along with a range of additional significant parameters, that determine photon behavior—here, size is not a decisive parameter for him. By affirming that photon and electronic behavior at a nano-object surface is not dependent on its scale (although he does not refute the role played by the size of the surface and of control that can be exercised on its physical properties), but on its structure and form, Borensztein introduces issues of materiality, and here epistemological questions immediately surface. He entertains the notion that an un-elucidated materiality of nanostructured gold particles exists and must be taken into account.

> Electronic intertwining and decrypting of those relations trace the optical effects that are the ultimate products: electrons at the surface of the particle will behave in such and such manner because of the form of the particle that tolerates them, and one sees this in the emitted light color. The relation occurring in this chain of relations lies between optical effects (emitted color) and the morphology of the particle... When the material possesses a specific form, it will yield a different effect. It is not a question of dimension, but rather of form... If the form of the surface particle is different, light will be absorbed in a different color, and this means that electrons vibration will change . . . they will resonate in green, in red, or in blue, . . . if the expected color of the surface is not yellow but, say blue . . . or even pink, this should indicate that the particle should be . . . flat or elongated. The light emitted and the color that this light has, indicates how the intertwined electrons act, oscillate, and vibrate around the particle's surface, and this gives information on the geometry of this surface.[53]

As we have already seen, the spatial dimension is fundamental to thinking and action in NSR. However, it would be imprecise to suggest that the design of properties is restricted to direct and exclusive consideration of form, claims Borensztein. The research of Borensztein and colleagues similarly takes into account elements such as plasmons,[54] and electron activities and orientation. Nevertheless this too is often connected with matters of form. While classical optics may describe some optical properties with reference to the facets of crystals, this is not sufficient to adequately project the performance of the nanosurfaces that are explored here. The scientist finds it difficult to formulate an effective descriptive vocabulary and to adopt terms such as "facets," "roughness," and "asperity."

[53] Interview of Yves Borensztein (INSP–"Physico-chimie des surfaces fonctionnelles") by authors (6 April 2008). Translated by authors.

[54] The plasmon is a quasi-particle resulting from the quantization of plasma oscillation (a superheated gas deprived of electrons). For example: Y. Borensztein, L. Delannoy, A. Djedidi, R.G. Barrera, C. Louis (2010) Monitoring of the plasmon resonance of gold nanoparticles in Au/TiO2 catalyst, *Journal of Physical Chemistry C*, 114(19): 9008–9021.

Within the geometry of the nanoparticles that Borensztein and his team examine, one must consider the relative spatial positions of atoms and molecules, their diameter, and height. A whole chain of reasoning and reflection emerges from this: a color of light is explained by the configuration of molecules at the surface of a nanoparticle (for example gold). From the study of this geometry, an explanation is given about the physical behavior of the electrons and plasmons engaged in the phenomenon. A step forward for Borensztein and his team is, through collaboration with chemists, to control the relative positions of the molecules deposited on the surface and in so doing, to organize the shape of this surface.

This knowledge allows the possibility of designing a material which can emit the wanted effect, by precisely controlling the position and spatial relations of the molecules on the surface.[55] In this perspective, materiality can be viewed as a kind of a loop beginning with high precision in deciphering the relations between the behavior of light color electrons and the surface shape of a nanoparticle, and finishing with the fabrication of such effects through the design of molecules on the surface of the material.

But "materiality" also resides here in a quest for parameters beyond those usually heeded in materials to be identified and to be included in research. This seems to be necessary in order to more fully appreciate the subtleties of some nanoproperties. This recommends an epistemology that calls for digging more deeply into nanomatter—a search for "hidden parameters." For example, in the view of Borensztein and colleagues, for reasons not fully elucidated, the inner structures of substances affect optical properties: entirely metallic solids exhibit properties that differ from objects whose core contrasts with the surface substance. Such complexification via a prism that incorporates the concept of the materiality of matter would ultimately reconfigure parts of the cognitive and epistemological landscape of nano-grounded questions. From the above analysis, to what does "materiality" refer? "Materiality" points to the host of surface and internal morphologies and interplaying forces that together lie within the architecture of a substance and that underpin the causes of physical behavior. In this instance, the epistemological tension is not between question/material nor material/question, but rather between a material and the crucial, active features which lie within the material. At the heart of Borensztein's research, the question is to identify which "materialities" figure, and how they operate and interact.

Indeed, the specificity and dynamics of a materiality and its localization often appear less important to researchers than the size and architecture of materials and their physical expression as observed as properties. In the complex and open-ended way that Borensztein deals with the materiality of matter, his research efforts introduce entirely novel epistemological issues concerning materials, the materiality of matter with reference to what counts as analysis, and as causes of physical properties. By raising issues of tension

[55] With other kinds of questions, Bernard Jusserand and Bernard Perrin's investigation on optical and acoustic waves in nanocavities can be understood in this same perspective (notably with the tailoring of their material and the introduction of Bragg mirrors for example) (see Chapters 2 and 6).

between dimension, form, surface effects, and internal structure, a new voice emerges that is not always present in reflections on matter or in discussions of relations between question and matter. Moreover, in this epistemological framework of "materiality," the very notion of control becomes highly problematized. Borensztein's investigations and reflections thus transcend standard thinking about matter and the tandem of question and matter by proposing that there exists and functions some still poorly understood domain, materiality, which expresses little understood fine-grain elements and operators, and which may require original forms of epistemology.

What does all of this imply, and how does it concern question-to-matter, matter-to question relations and broader epistemological issues? Borensztein's line of reasoning suggests that parameters beyond those usually heeded in materials must be identified and must figure in research. This seems to be necessary in order to more fully appreciate the subtleties of some nanoproperties. Such complexification via a prism that incorporates the concept of the materiality of matter would ultimately reconfigure parts of the cognitive and epistemological landscape of nano-grounded questions. In this scenario, what counts as cause and how cause operates are re-defined. This line of investigation and thinking is rooted in an epistemology of form where form is itself explained through a highly fine-grained analysis of its effects and where effects are themselves understood through the detailed comprehension of the behavior of elements engaged in the linking of causes and properties. In this way, the habitually inflexible notion of well-defined, highly characterized, and stabilized matter is replaced by a plastic descriptivism. Here, both question and material dissolve into inquiry about precisely which features exist and operate inside a material, and how these features interact. The whole of this signals a central point: the above broached tandem of the question-to-matter and matter-to-question spiral has here, through the epistemology of materiality, been importantly extended and rendered more complex.

4.4 THE EPISTEMOLOGY OF NANOBIOLOGY RECOGNITION

One can identify certain fields of nanoscale research where numerous aspects of epistemology occur and converge. Such cases illustrate the roles of epistemological components and how they interact. Nanobiology affords the possibility of appreciating nanoepistemology at close range and in dynamical operation.

In the field of nanobiology, processes of biological recognition are pivotal. Recognition frames aspects of reflection acting as a confederative mechanism. The concept of recognition indeed introduces sets of epistemological markers that constitute a roadmap, which mobilizes the understanding of processes, and ultimately nourishes explanation. Precisely, why is recognition so central to nanoscale research epistemology in biology? As we have seen in Chapter 3, one of the elements of the NSR hexagon is "binding." In the context of nanobiology, binding means both chemical affinity and form adjustment.

When a complex molecule like a protein or an enzyme encounters and reacts with another molecule, it must meet the proper site to fix; this entails not only chemical bonding but an adjustment (in terms of spatial occupancy and shape) between the active sites of the two molecules. The same process describes the way a molecule folds and connects some of its sites to other parts of itself.

Historically, the image of recognition in terms of a lock and key has figured importantly in biochemistry, and it has pervaded reflection.[56] The special status of recognition in the epistemology of nanobiology is due to the fact that binding is no longer envisaged in terms of a static match between lock and key, but is now instead understood as a dynamical event where the key is seen to turn in the lock. Recognition is here understood as a process in which complementary molecules orient themselves in order to fit and fix to their partner molecule. This now means dynamics and temporality have deeply penetrated reflection. Dynamics here becomes an epistemology of nanobiology.

With the advent of nanobiology and thanks to instrumentation that permits observation through images, the era of lock and key as a simple metaphor for biomolecule connections has passed. The images provided by AFM, for example, give access to the possibility of "seeing" the local adjustment between three-dimensional biomolecular entities. One can now describe the morphology of the target site and of the complementary element. As discussed in Chapter 3, a horseshoe-shaped protein whose inner surface consists of numerous specifically configured entities, recognizes corresponding molecules, and captures them (see Figure 3.8).[57]

In the epistemology of nanobiology, recognition functions as a nucleus to which other elements gravitate and federate. This capacity to reason in a framework of form is expressed by spatial parameters: considerations of molecular orientation, twisting and binding, trajectory, etc. This means reasoning in complex multi-dimensional space and in terms of travel, which considerably enriches the epistemology. This is nowhere better seen than in the status of three-dimensional representation today so prevalent in nanobiology. Epistemological clusters emerge in this case with outstanding clarity. Here we see an example of federation around recognition, so central to reflection in nanobiology.[58] The combination of consideration of recognition plus form, plus space, plus three-dimensionality, plus travel, is central to the understanding of biomolecular bindings, recognition, and function. It is through this combination that the dynamics of entities such as the "legos," with well-defined shapes and properties, can be profoundly understood. Moreover, as shown in Chapter 3, it allows reflection to penetrate more and more sharply into extremely complex questions, such as those about the evolution of life. This epistemological ensemble constitutes a fully fledged expression of descriptivism through which a deep comprehension of molecular processes inside bioentities can be reached.

56 Cl. Debru (1983) *L'Esprit des Protéines*. Paris: Hermann.

57 Interview of Philippe Minard (IBBM, University of Orsay) by authors, 12 June 2010; e.mail communications from Philippe Minard to authors (3 June 2013, and 3 November 2013).

58 Interview of Philippe Minard (IBBM, University of Orsay) by authors, 12 June 2010.

Concerning the place of sequencing in nanobiology, its epistemological force and originality lie in the complex relation between the sequence, the resulting form of biomolecules and, consequently, their recognition potential. The idea of sequence is foundational to the reasoning and understanding associated with biological molecules. The novelty of nano in this epistemological tandem (sequencing and recognition) consists of the introduction of sequence through the lens of form.

Sequencing entails the idea of arrangement, order of a limited number of items (the basis for nucleic acids and amino acids in proteins), in an intended or, if unintended, a known configuration, which is the foundation of recognition. There is always a restricted number of elements, organized according to a precise hierarchy, that combine in myriad different fashions to construct the biomolecule as the recognition molecule or as the target molecule. These sequences, be they triplets of bases or amino acids, are read in the framework of form. They are understood as form. The epistemology of form is entirely inescapable here. The notion of arrangement is transverse; it applies equally to sequence and to form. For sequencing, the nano-epistemological shorthand is synonymous with arrangement. This concerns the understanding of how a biological molecule like a peptide is built, but also the possibility of creating novel molecules. In nanobiology, control is a central concern. By selecting the sequence in the DNA, one can shape the configuration of the protein components. The triad sequencing/arrangement/form opens the possibility in nanobiology to think of functions (among which recognition is paramount) and properties of biological molecules in a renewed perspective. This is a clear expression of the place of determinism in nanobiology.

CONCLUSION: THEORY, MODELS, AND DESCRIPTION

In the following pages, we first locate NSR with reference to some philosophy of science literature dealing with the structure of theory, and more particularly the structure and functions of models, notably as set forth by Margaret Morgan and Mary Morrison in their concept of mediating models. This is followed by reflection on what a model might mean in NSR. These thoughts then lead us on to discuss the above identified component of the epistemological framework of NSR. The chapter closes with some general reflections on how an understanding of work in nanoscale research may cast light on some important transformations of theme and epistemology from parts of early-twenty-first-century scientific investigations.

Now for some reflections that have been formulated by philosophers of science on theory and models, During the opening decades of the twentieth century, much philosophy of science was synonymous with discussion of theory. It was held to be the driving force behind knowledge and the guiding framework for experimentation. What is known as logical empiricism, the "received view" prevailed. Theory was understood to be syntactic in composition; that is to say logically comprehensible in terms of the language expressed therein, and in the logical progression of the reasoning that it carries. It consisted of two categories of statements: theoretical sentences and observation sentences. The meaning of theoretical statements was given by observation sentences, and since this proved extremely problematic, what were known as correspondence rules were introduced to bridge the gap between observation and theory. The narrowly logical excesses of the received view of logical empiricism became subject to criticism in the middle and later decades of the twentieth century, when a semantics-grounded perspective of theory emerged.[59] The syntactic perspective deciphers the logic which links the different words in a key sentence in a scientific theory, whereas the received view had no focus on the meaning and content of a scientific theory. The semantic perspective introduced meaning into considerations of theory. However, this perspective is certainly not congruent with what observers of the scientific practice of research would recognize, nor could it be acknowledged by research practitioners.[60] The degree to which NSR epistemology converges with such philosophy of science perceptions of theory will now be examined.

Like the philosophical perspectives of theory (syntactic and then semantic), the understanding of what constitutes a model, and how models operate, has also evolved along two lines. Many philosophers of science initially regarded models as auxiliaries to theory.[61] Investigation of the epistemological place of models by philosophers of science long concentrated on the relationship between models and theory. Models were notably held to be auxiliaries of theory. They served to further clarify theory statements. They constituted idealizations that could enlighten theory. Alternatively, tests of the validity of a theory could be examined in terms of its consistency with the theory's model. It must also be recognized that in this view, models constituted a vague connection between experimentation-based claims and theory.[62]

Since the 1990s,[63] a refreshing and penetrating perspective on models has emerged, in large part connected with the work of Mary Morgan and Margaret Morrison. These two philosophers of science introduced the concept of models as mediators.[64] In their view, models are autonomous. They cannot be understood either as a product of theory or as driven by experimentation. They belong to neither referent, and this is the source of their

[59] F. Suppe (1977) *The Structure of Scientific Theories*. Urbana, IL: University of Illinois Press; F. Suppe (1989) *The Semantic Conception of Theories and Scientific Realism*. Urbana, IL: University of Illinois Press; P. Jacob (1980) *De Vienne à Cambridge*. Paris: Gallimard.

[60] B. Latour, S. Woolgar (1986) *Laboratory Life: The construction of scientific facts*. Princeton, NJ: Princeton University Press.

[61] N. Cartwright (1983) *How the Laws of Physics Lie*. Oxford: Clarendon Press; N. Cartwright (1991) Can wholism reconcile the inaccuracy of theory with the accuracy of prediction? *Synthese*, 89(1): 3–13; N. Cartwright (1997) Where do laws of nature come from? *Dialectica*, 51(1): 65–78.

[62] M. Black (1962) *Models and Archetypes. Models and metaphors, Studies in language and philosophy*. Ithaca, NY: Cornell University Press; M.B. Hesse (1966) *Models and Analogies in Science*. Notre Dame, IN: University of Notre Dame Press.

[63] Interestingly, the topic of models has long been of concern to practicing scientists. For example, Hertz and Maxwell reflected on the role they play in the organization and representation of experimental findings.

[64] M.S. Morgan, M. Morrison (eds.) (1999) *Models as Mediators: Perspectives on natural and social science*. Cambridge: Cambridge University Press.

autonomy. Theory, experiment, and model thus form a kind of triangle. An important characteristic of mediating models is their capacity to converge and organize pure data and theory. They contribute to intelligibility in both: they mediate between them. Morgan and Morrison argue that models constitute intellectual instruments. It is in part in their capacity as instruments that, with reference to data, models can be used to filter and structure information or to participate in the design of experiments. Models are often regarded as being representative of a family of properties or even several phenomena, as opposed to representations associated with a single experiment, property, or family of material. In this sense, models must be viewed as having some affinity with medium-level generalization. Beyond this, as expressed by these two philosophers of science, mediating models exhibit two additional key features: (1) mediating models provide the criteria for refinement of theoretical description; (2) models can themselves become the objects of interrogation, as opposed to promoting interrogation about the structures and behavior of the physical world.[65] Another feature revolves around the key issue of simplification, which is considered as central by most contributors to Morgan and Morrison's book, *Models as Mediators*. The simplification entailed in modeling is requisite on selection. A model without simplifications would be so encumbered and complex that it could not operate as a model. Because models are intended to represent objects from a diversity of phenomena, they are necessarily constructed in terms of what are considered as relevant choices. As will be seen, this issue of simplification, in terms of relative absence, is a cornerstone of NSR epistemology.

Where is the place for theory in NSR? In what context does it occur and to what purpose? The term theory covers a lot of territory, and one must establish exactly what part of that territory in NSR is referred to by the term theory. Theory as an object of research or as a specified resource does not lie at the forefront, either in computational or metrological experimentation. Our interviews of dozens of scientists working in nanophysics and nanobiology fail to exhibit involvement in the construction or modification of theory, nor do the results of these scientists directly integrate defined and specific theoretical claims. There are no signs of theoretical novelty or ambition. Theory is mainly essential as background knowledge, which is also the case with science at large. Stated differently, theory is employed by scientists, but in NSR, new theory is rarely produced. Particularly in metrological experimentation, theory is not directly a component of the production of description. Its role is instead passive. On a different register, however, theory is central to simulation experiments. Theory, along with semi-empirical components, enjoys a place in the algorithms used in simulation, but it is a silent partner. It is not explicitly mobilized by simulators when they discuss their research operations or their findings. However, and this is an important point, the crucial work of prediction performed by simulation is indirectly steeped in theory, and it expresses two important epistemological expectations of theory, namely anticipation and explanation. It is the predictions and explanations offered by simulation that nourish reflection on the possible links that might exist between the metrological

[65] R. Giere (2010) Explaining Science: A cognitive approach.

findings occurring in a specific class of materials or with reference to a category of properties. It is in the tension between metrological and simulation findings and representations that a glimmer of theory appears. An instance of this is documented in Chapter 2, in our description of the "gray story," in which metrological and simulation experiments converge. It is in this oblique sense that theory figures in NSR and not in the more conventional sense of abstraction, generalization, and integration that are commonly associated with the function of theory in many other scientific domains. In a word, theory in the strong sense of the term is absent; only theory as a remote and unspecified category is present. One can speak here of unintrusive theory only. Why this is so, will be discussed below.

The presence of mediating models is similarly rare. By their very definition, mediating models form a triangle between theory, model, and empirical data. Because of the very absence of theory-related questions or reflections, models in the form of mediating models are not an epistemological option. Mediating models are absent from NSR for an additional reason. Their function is to integrate and represent information drawn from multiple experiments and related to multiple properties or types of phenomena or materials. The notion of integration is foremost here. In all instances, the model is in some respect an idealization of phenomena—a selective simplification expressed as an idealization. Such simplifications and reference to multi-class representations are alien to most NSR thinking and representations. Mediated models strive for some level of generalization. NSR rarely generates generalization. To the extent that models occur at all in NSR, they do not express simplification and generalization: they are not integrating. The models of NSR instead revolve around the specific. It is diversity and considerations of the local that prevail. In view of what they depict, describe, and explain, the models of NSR are best characterized as "local models."

If NSR does not privilege theory and models, what then is the epistemological grammar and focus of the field? Throughout this book, we have argued that NSR is essentially a descriptive science. The term "descriptivism" has sometimes been used to characterize certain research results from the end of the nineteenth century, particularly in areas of physics. Historical descriptivism was grounded on two considerations. The first focuses on accuracy of measurement. For example, in spectroscopy, wavelength was worked out to the decimal point. Crystals were studied in the light of their numbers of facets and the angles between them. Second, understanding of objects was linked to an appreciation of their form.

Description constitutes a key epistemological logic and language of NSR. The spatial profile of single objects and the spatial relations between clusters of objects lie at the heart of description. Spatial relations are identified through information-yielding images. But in NSR, images are more than information, since they also constitute a foundational component of reasoning and are in this sense crucial to epistemology. Images yielded by instrumentation reveal form, and it is form that is characteristically reported in the published description of objects. To describe an object's form is in some important ways to understand the object, and in other instances to grasp its properties and behavior. Typical terms of description include words such as: bridge, trough, island, box, wire, tube, armchair, crenellations, curve, edge, hexagon, V-shape, network, vortex, etc. Such expressions assuredly do not function here as metaphors. They instead rigorously refer to the geometries, topologies, and curvilinear configurations visible in the images of metrological or numerical experimentation images. These terms by and large operate as the medium of description of objects and phenomena which revolve around form, which as indicated throughout this book, comprises the federative component of the combinatorial epistemology of NSR. Finally, in NSR, description is situated in a local setting. The setting is defined by a target event and its immediate environment, which may consist either of other events or a backdrop (such as a substrate). In all cases, NSR defines local as nanometric in dimension. In nanoscience, the aim is not to see broadly; it is not to move toward the general.

Descriptivism is predicated on a deterministic perception of physical systems, and to this extent, it too figures importantly in the combinatorials of NSR epistemology. In NSR, the existence and centrality of form flows from determinism. Form may be regarded as a signature of determinism, and determinism allows one to know the relative positions of elements, such as atoms, and in the same light to exactingly predict the form of the surface. Determinism reflects the capacity to grasp unit-by-unit an ensemble of elements. The expression unit-by-unit refers to stable discrete units which fit together to constitute a coherent collective. In NSR, determinism operates successfully at the atomic and molecular levels. Scientists observe single molecules, and it is rumored that some practitioners can observe molecular oscillation.[66] The epistemological implications of determinism are vast and are central to the domain of NSR.

Determinism is also indeed expressed as control on the atomic and molecular levels. The requirements entailed in NSR control cannot be satisfied by collective behavior and by probabilistic motion. Each atom and each molecule has its particular place. It is the task of NSR practitioners to know this place, to be able to predict it, and to be able to manipulate it. Control is expressed in a multitude of ways. It is omnipresent in the fashioning of synthetic materials, and in the even more precise manipulations required by research-objects and materials-by-design called for in many experiments. Control of substances is used actively in the course of experiments in order to determine the physical consequences of intra-phenomenon transformations. In such instances, control is involved in both producing and studying dynamics. This is notably the case in nanobiology (see Chapter 3). In NSR, the meaning of control stretches beyond engineering. In the latter, it constitutes a means to an end. In nanoscale scientific research, control is part and parcel of the epistemology of cognition itself, because of its place in materials' experimental design and the exploration of internal processes of phenomena.

In NSR the goal is certainly not the collection and assimilation of information into a model; on the contrary, it is the preservation of the singularity of each stable result and a comparison between like and unlike findings. Knowledge enhancement is often generated through comparison, and comparison of descriptions is here so crucial that, among its many other designations, NSR may be viewed as a science of comparison.

[66] Interview of Mark Ratner (North western University) by Terry Shinn, 29 January 2008.

5

The Role of Combinatorials in Structuring NSR Cognitive Trajectories

Nanoscale scientific research is the kingdom of combinatorials, and their profusion powerfully affects the cognitive trajectory of many nanopractitioners. In the following, we will show how combinatorials are played out in practices that we refer to as "respiration." The research horizons offered by combinatorials are vast. Respirations make possible a deepening of an established research line; and they alternatively permit the interruption of research lines and projections into a very different domain. In this perspective, through respiration, combinatorials may be seen as resources that open windows onto new opportunities. The very abundance of combinatorial resources frequently induces NSR practitioners to pause in the course of their investigations, and in the light of the possibilities that they provide, to rethink their research trajectories. This combinatorial-based respiration takes the form of a kind of interval that either reaffirms a new start in the existing research project, or initiates the launch of a new research orientation. NSR Scientists interrupt their activities and take a breath before resuming work. This momentary pause is respiration.

Respiration involves two different spheres of activity. The selection of specific sets of combinatorials by practitioners generates "epistemological expectations," which occur in what we term the "concentration sphere." Alternatively, opting for different combinatorials induces "function-horizons," and this occurs in the "extension sphere." Inside the concentration sphere, NSR scientists decide during intervals of respiration to regroup their forces in an established project through recruiting additional combinatorials; or they instead opt to shift to a new intra-concentration research arena. On a second level, practitioners may choose to shift from the concentration sphere to the extension sphere, and this once again entails a deep breath in the form of a respiration. We note that, when practitioners move from the concentration sphere to the extension sphere, two respirations frequently occur, as they generally return at some point to their previous sphere.

Circulation within and between concentration and extension is punctuated by a crucial brief temporal interval of respiration, when scientists assess their current cognitive situation in terms of their fields of curiosity, the various conceptual and material resources they have previously mobilized, the emergence of new intellectual and investigatory resources, and the possible merit of acquiring a fresh set of components to be added to those already existing, for adventures in a new arena in the same sphere, or in view of passage into the alternative sphere.[1] We examine what occurs in NSR during respiration intervals and suggest that respiration offers key insights into cognitive and social operations.

In the first section of this chapter, "The respiration model," we set forth the framework of respiration in NSR. The factors that propel practitioners out of a research arena into a respiration interval occurring between arenas in the same sphere, or that propel them into respiration associated with a change to another sphere, are identified, and the concerns and considerations that practitioners experience in the respiration space are described. A vocabulary for characterization of the internal cognitive operations that occur within concentration and within extension are introduced, and we show how circulation inside and between the spheres is grounded in respiration intervals.

How respiration unfolds in NSR, with reference to combinatorials in the concentration sphere, and documentation about precise epistemic expectation dynamics, are discussed in the second section, "The turf of concentration." The third section, "Gifts of extension," mobilizes a second NSR case study to describe the workings of movement from concentration to extension and then back to the concentration sphere, in an NSR context. The very issue here consists of pauses where practitioners are temporarily suspended before they opt for a selected path. It is once again demonstrated that combinatorial-based considerations are of the foremost importance.

5.1 THE RESPIRATION MODEL[2]

In this section we discuss the substance and functions of what we term "respiration," sketch the orientation and priority of the concentration sphere and the extension sphere,

[1] The question of intellectual mobility has received only limited attention in the history and sociology of science. Sedentary and nomadic trajectories have been explored for molecular biology in a psychological framework. Nomadism is defined as practitioner involvement in a succession of alternative research domains, versus sedentary which refers to thematic stability; see A. Marcovich (1976) La Créativité des hommes de sciences, Diplome de l'École des Hautes Études en Sciences Sociales, Paris. Degrees of intellectual mobility have also been analyzed with reference to practitioners' utilization of their favored instrument for resolution of problems in multiple fields; cf. M. Mulkay (1974) Conceptual displacement and migration in science: A preparatory paper, *Science Studies,* 4: 205–234; The carriers of Georges Gamow and Max Delbrück, indicate that the promise of innovative cognitive innovations also sometimes motivates changes (L.E. Kay, 2000) *Who Wrote the Book of Life?: A history of the genetic code.* Stanford, CA: Stanford University Press; N. Mullins (1972) The development of a scientific specialty: The phage group and the origins of molecular biology, *Minerva,* 10(1): 51–82). Finally, mobility occurs as a "bandwagon effect," where scientists react to professional and financial inducements (J.H. Fujimura 1992) Crafting science: Standardized packages, boundary objects and "translation." In: A. Pickering (ed.) *Science as Practice and Culture.* Chicago, IL: University of Chicago Press; J. H. Fujimura, M. Fortun (1996) Constructing knowledge across social worlds: The case of DNA sequence databases in molecular biology. In: L. Nader (ed.) *Naked Science. Anthropological inquiry into boundary, power and knowledge*. London: Routledge).

[2] A. Marcovich, T. Shinn (2013) Respiration and cognitive synergy: Circulation in and between scientific research spheres, *Minerva,* 1(51) .

and indicate how respiration works with reference to the two aforementioned spheres.[3] Respiration may be defined as a temporality, a relatively brief "interval" possessing two specific characteristics: reflection and combinatorials.

Respiration in research activity consists of those in-between project intervals in a scientific trajectory, when a scientist takes stock of present and likely future intellectual possibilities, and considers ensuing research options and the best path in view of mainly cognitive criteria. It corresponds with the precise interval that separates involvement in one research project from engagement in a same or alternative project. It is a period of suspension between the consciousness of present constraints and combinatorials, and the calculated potential of new reference and combinatorial resources. The significance of combinatorials in NSR makes discussion of respiration unavoidable.

It is bracketed by "research arenas," which are long periods when practitioners operationalize decisions and initiatives formulated during respiration, and where they generate research findings. Respiration is then a period when the daily absorption in research activity is replaced by reflection—perhaps reflexivity[4]—and prospect. Respiration constitutes an opportunity for openness to new possibilities of combinatorials in NSR, and it also supplies insights useful for research stability. In so doing, it is a key element within the processes of NSR. It is perhaps not excessive to suggest that respiration constitutes the in between, reflective, stocktaking window for the birth, consolidation, adaptation, or even redirecting of much NSR endeavor. Reflection thus constitutes the first defining characteristic of respiration.

The substance and dynamics of combinatorials themselves comprise a second aspect, indeed the very cement of respiration. It is probably useful to remind the reader here that combinatorials include items like scientific concepts, research instruments, materials, skills, and audience. They are selected by scientists for association and integration in order to address and resolve an original research question or to attain an objective. As shown in Chapter 2, an instance of a new combinatorial is exemplified by the association of Raman spectroscopy (for optics detection) with Bragg mirrors (for acoustic detection), and a novel semiconductor with the production of nanocavities.[5] Assessment of the strengths and coherence of combinatorials for a candidate research project thus lies

[3] The model of science and technology research regimes maps poorly onto the model of respiration. The transfer of practitioners from a disciplinary regime to some alternative regime entails the acquisition of another cognitive set and of a different ensemble of expectations. It also entails a switch of audience. This occurs on what must be considered a grand, macroscopic level. In contrast, the respiration model focuses on the actions of individual practitioners, and its force lies in the capacity to illuminate sometimes settled deviations in topic, question, instrumentation, or material. Respiration may be regarded as an extension and refinement of discussion concerning sedentarism and nomadism in scientific research (B. Joerges, T. Shinn, 2002). *Instrumentation between Science, State and Industry.* Dordricht: Springer; T. Shinn (2008) *Research-Technology and Cultural Change: Instrumentation, genericity, transversality.* Oxford: Bardwell.

[4] P. Bourdieu (1975) La Spécificité du champ scientifique et les conditions sociales du progrès de la raison, *Science et structure sociale*, 7(1): 91–118; P. Bourdieu (2001) *Science de la Science et Réflexivité*. Paris: Raisons d'Agir.

[5] Interviews with Bernard Perrin and Bernard Jusserand (INSP University, Paris) by authors, 29 November 2007 and 28 June 2008.

at the heart of the respiration interval. Selection of specific elements for the combinatorial frame, the manner of their intertwining, the priority and hierarchy of items, and the purpose to which they are put are central ingredients. A strong combinatorial package may itself become a kind of novel combinatorial.

As we have observed within NSR, respiration is most fundamentally a process model. The underlying meaning of the several components associated with respiration is grounded in process and dynamics. In respiration the above-indicated ingredient of reflection occurs gradually over time and transforms from initial musings to analytic calculation to mature commitment. Combinatorials likewise constitute a process where elements are disassembled, re-assembled, and then the cycle begins once again inside a single respiration interval. The circulation linked to respiration is a process, sometimes linear and in other instances cyclical. Finally, long-term and short-term temporality (to be introduced next) entails processes that have an internal movement inside arenas or more generally between science spheres.

Respiration is significant for NSR because it contributes to cognitive synergy inside the concentration sphere and the extension sphere, and to the dynamics of the back-and-forth circulation between the two spheres. What are the main features of concentration? How do they operate? What is the place of respiration? As observed in NSR, concentration refers to a research space characterized by convergence between ideas, instruments, materials, skills, and also by communities of scientists with reference to a specific project, and most significantly governed by dominantly epistemic expectations. The notion of epistemic expectation refers to the intentions of practitioners to study categories of matter, largely in order to describe its properties and to explain the organization and operation of relevant forces, where such objectives (knowledge) are regarded as sufficient and important in their own right (Chapters 1, 2, and 3), and thus to a great extent only remotely coupled to exogenous inputs and interests. In a word, in concentration everything is funneled toward epistemic expectations.

The epistemic expectations of the concentration sphere are organized mainly around scientific disciplines. Nevertheless convergence in the efforts of practitioners from neighboring disciplines is sometimes evoked, and such synergistic collaborations constitute in themselves an expression of the combinatorial matrix. Yet the disciplinary referent remains dominant, providing the central question of the fields, methodology, and appropriate guidelines of rigor. Exchange with other disciplines acts as a supplementary resource for the identification and assembly of combinatorials (see Chapter 6).

Finally, the concentration sphere is associated with long-term temporality. This contrasts with the short-term temporality of the extension sphere. In concentration, brief intervals of respiration occur between long periods of involvement in a project. Respiration thus demarcates extended periods spent in one intra-concentration research arena, before the practitioner circulates to a different intra-concentration arena, or alternatively, adjusts the existing arena through injecting fresh combinatorials. The long temporality of concentration accords with the preponderance of disciplinarity, and the presence of respiration intervals can provide a space for creative adjustments to sometimes over-defined disciplinary frames, through offering a window for the legitimate entry of novel components.

The structures and dynamics of the extension sphere differ from the concentration sphere. In extension, practitioners operate in a framework of "function-horizon." The term "function" refers to the study of materials in terms of their capacity to achieve an outcome and not in terms of an understanding of the structures and forces of a material per se. This centrality of function in extension does not in any way mitigate the cognitive, scientific aspect of the research endeavor. Two approaches to research on electrical conductivity in germanium on the morrow of World War Two are illustrative. While working at AT&T (the American Telephone and Telegraph Company) in 1947, John Bardeen and William Shockley (both winners of the 1956 Nobel Prize in physics) studied germanium crystals, mainly from the perspective of semiconductor physics theory;[6] by contrast, at almost the same time, scientists working at RCA (Radio Corporation of America) saw germanium exclusively in terms of an amplification function.[7] Included in the notion of function is the stability, reliability, and efficiency of an artifact and its utility for designated purposes and publics. In NSR, the scientists whom we observed in our study of the extension sphere originated their trajectory in the concentration sphere. The cognitive concern of scientists with function inside extension may be seen as a wish to express concentration-rooted research findings and insights in a multifarious, extra epistemology expectation framework. The concept of expression is central to extension. It signals the desire to translate epistemic claims into a variety of forms and formats. This is not the erstwhile stuff of the often so-called linear perspective "endless frontier" of the 1940s and 1950s;[8] rather it is the curiosity of a scientist to see his research results played out in a variety of material domains and perhaps social environments. This may be regarded as a characteristic of NSR.

This introduces an additional central theme—that of extra disciplinary audiences and horizons. The term "horizon" refers to the open-endedness and scope of possibilities relevant to the different audiences. "Function-horizon" then deals with control and standardization, on the one hand, and with organization, economy, regulation, government, and market, on the other.[9] If one adopts the concentration/extension perspective, the extension sphere must be regarded as "open" because of its "inclusionary" features, and concentration must be viewed as closed, because of its funneling features. This observation raises questions about some current representations of research organization, which associate academic work as open, versus entrepreneurial work as private or closed. At least in the case of NSR, function-horizon constitutes a territory where scientists are free

[6] M. Riordan, L. Hoddeson (1997) *Crystal Fire: The birth of the information age.* New York, NY: WW Norton & Company.

[7] H. Choi (2007) The boundaries of industrial research: Making transistors at RCA, 1948–1960, *Technology and Culture*, 48(4): 758–782.

[8] V. Bush (1945) *The Endless Frontier*. A report to the President by Director of the Office of Scientific Research and Development. Washington, DC: United States Government Printing Office.

[9] A. Marcovich, T. Shinn (2011d) From the triple helix to a quadruple helix? The case of Dip-Pen nanolithography, *Minerva*, (2):175–190.

to communicate, exchange, and share with different audiences, yet not at the expense of cognitive attainments.

A difference between extension and concentration also resides in aspects of their combinatorials. In concentration, the same categories are always involved (ideas, instruments, materials, skills), while in the extension sphere the categories of elements entering into combinatorials vary from case to case and are very context dependent. In function-horizon, the quantity and diversity of considerations relevant to the production of results is vast. In extension, they are marked by a variety of logics, for example technical, administrative, public health, ethical, market, manufacturing, political, and cognitive. All of these features constitute combinatorials in this sphere that are connected to the substance of function-related cognition. Conversely, in concentration, one perceives a kind of "defined equation" of combinatorials, sharply focusing on an epistemic-founded question, versus the dominance of very open-ended and contingent combinatorials in extension. In a word, concentration is synonymous with "curiosity," whereas extension opens on "adventure."

Finally, respiration, concentration, and extension are connected by different configurations of circulation. The point of departure of circulation resides in respiration. Inside concentration, circulation is expressed as the movement of a practitioner from an intra-concentration arena to a different intra-concentration arena, based on assessments occurring during respiration. In a second configuration, scientists switch from the concentration sphere to the extension sphere. Lastly, inside the extension sphere, the aforementioned scientist returns to the concentration sphere. One may attribute circulation back into the concentration sphere as an expression of the power of epistemic expectation and the disciplinary referent. In Sections 5.2 and 5.3, we will employ case studies to document the research processes occurring inside the concentration and extension spheres and will examine the place of the respiration interval as a prelude to circulation between spheres or maintenance within a sphere.

5.2 THE TURF OF CONCENTRATION

The cognitive trajectory of Shimon Weiss, an Israeli-born nanoscience inclined physicist who today directs a large laboratory at the University of California, Los Angeles,[10] is emblematic of research activities in the concentration sphere, which are associated with two temporal intervals of respiration—one in *c.*1995–8 and the other in *c.*2006–7. It is to be noted that questions of combinatorials lay at the heart of both respiration moments.

Between 1986 and 1995, Weiss conducted investigations using a largely stabilized and coherent package of combinatorials in a tranquil research arena demarcated by issues of optical and laser spectroscopy, signal mixing, and semiconductors. This was followed by

[10] Department of Chemistry and Biochemistry, Department of Physiology, University of California, Los Angeles.

an interval of respiration of about three years, where two sets of original combinatorials were put in place. In this respiration, Weiss compared the possibilities of new combinatorials with past achievements and evaluated their potential contributions. Here he introduced single molecules (inorganic then biological molecules), a new marker system, and biocompatible quantum dots in the framework of the spectroscopy orientation—the latter continuing on from his earlier physics research. These investigations clearly placed Weiss in the nanoscale research framework. This gave rise to a second long-term concentration research arena (1998–2006) where a stable, professionally acknowledged disciplinary community shares well-defined epistemic expectations around specific questions or techniques. In a second respiration episode, beginning in around 2007, he introduced additional novel combinatorials. Previously, biological molecules had simply been material for the optical study of physical phenomena. In this second respiration, however, he re-positioned his own vision of biological molecules, which became interesting in themselves. In a now third concentration arena, in collaboration with biologists, Weiss began to study the way these molecules behave, how they initiate and terminate their replication process, and how they mute. Here Weiss contemplates questions of disease and treatment, and this emerges as one important combinatorial.

5.2.1 From a first concentration arena to an initial respiration episode

We will now describe Weiss's orientation during his first concentration episode, which extended over almost a decade. Between 1986 and 1995, Weiss mobilized a number of complementary combinatorials that focused principally on matters of optics and semiconductors. Much of this work was carried out in the Bell Laboratories.[11] Near to the end of this period, his research moved into the direction of quantum and nanoscale research. In his early days, Weiss addressed problems relating to wave mixing, where he used optical lasers. He soon extended this work into the area of photodiffracting lasers, and not long after that he introduced consideration of semiconductor materials. It would be a mistake to think that Weiss's investigations were entirely devoid of theoretical and experimental aspects, as indicated in one of his early articles, entitled "Theoretical and experimental study of frequency detuning in photorefractive oscillators."[12] By 1993–4 Weiss's attention had turned to the observation of excitons,[13] which reinforced his use of semiconductors.

[11] The Bell Laboratories are located in industrial settings. But there is no necessary linkage between organizational setting and the kind of research carried out in laboratories. For many decades after World War Two, Bell employed up to 1000 scientists, who were allowed total liberty in their selection of research projects. Weiss's engagement in concentration reflects the epistemic expectations openness of Bell at this period of its history, when there existed almost no pressure to engage in extension activity.

[12] B. Fischer, S. Sternklar, S. Weiss (1986) Theoretical and experimental study of frequency detuning in photorefractive oscillators, *Journal of the Optical Society of America A-Optics Image Science and Vision*, 3(13): 34.

[13] An exciton is a quasi-particle that is found in semiconducting materials. It is a pair formed by an excited electron that has acquired a higher energy level and the resulting "hole." When the excited electron recombines with its hole, energy is generated, which takes the form of a photon emission; L. Apker, E. Taft (1950) Evidence for exciton-induced photoelectric emission from f-centers in alkali halides, *Science*, 112(2911): 421.

It was during this period that he added to his already considerable arsenal of cognitive elements that would open the way to the 1995 respiration interval. The range of items consisted of quantum dots and development of new nanometric observation instruments in the guise of scanning tunneling microscopy (STM) and near-field optical microscopy (NFOM).[14] A key constant underpinned all the work in this optical exploration arena where the whole of his work revolved around observation and analysis in the framework of images.[15] Weiss's initial respiration moment, which covered the period 1995 to 1997, opened with an assessment of future research possibilities offered by his recent achievement of the observation of a single molecule. This was accomplished through construction of the aforementioned NFOM, which was initially intended for the study of excitons. In the course of using this device, Weiss noted that he could also discern single molecules. This exciting discovery led him to envisage the study of interactions between two neighboring nanoscopic molecules through optical signals. In order to study these interactions, he adopted an existing marker technique, the Förster resonance electron transfer (later called fluorescent resonance electron transfer: FRET).[16] The shift toward an interest in single nanostructured molecules and chemical markers comprised the introduction of two new combinatorials to the existing earlier package of optical laser observations and semiconductors (both of these combinatorials reoccurring importantly in subsequent investigations).

His ability to observe single molecules prompted him to ask questions about molecular behavior when two molecules are in close proximity—one becoming the environment of the other. Using FRET and for purposes of observation, he needed to tag the two molecules. However, Weiss soon learned that the investigation of single molecules using FRET had long been a topic of much research. In response to this, the young physicist was challenged to develop techniques for the tagging of very specific locations within a single molecule. This implicitly entailed a high degree of control.[17] At this time, *c.*1995, Weiss found himself in a period of deep questioning, so characteristic of a respiration moment. He soon came to realize that DNA molecules offer an extraordinary potential for controlled marking. By the mid 1990s, synthesis of DNA was commonplace. Weiss could thus readily control specific structures and locations inside DNA molecules. This appropriation of DNA occurred in the framework of requirements for the kind of tagging often employed in physics research on inorganic substances. For Weiss, DNA molecules were merely a convenient platform. At this time, the biological functions of DNA

[14] S. Weiss, D.F. Ogletree, D. Botkin, M. Salmeron, D.S. Chemla (1993) Ultrafast scanning probe microscopy, *Applied Physics Letter*, 63(18): 2567–2569.

[15] Interview of Shimon Weiss, professor at UCLA and the Maison des Sciences de l'Homme Paris, by authors, 22 June 2008.

[16] Förster (Fluorescence) resonance energy transfer (FRET) is a mechanism describing energy transfer between two chromophores. A donor chromophore, initially in its electronic excited state, may transfer energy to an acceptor chromophore through nonradiative dipole–dipole coupling. <http://en.wikipedia.org/wiki/F%C3%B6rster_resonance_energy_transfer> (consulted 3 May 2012).

[17] T. Ha, D.S. Chemla, T. Enderle, S. Weiss (1997) Single molecule spectroscopy with automated positioning, *Applied Physics Letters*, 70(6): 782–784. T. Ha, T. Enderle, P.R. Selvin, D.S. Chemla, S. Weiss (1997) Quantum jumps of single molecules at room temperature, *Chemical Physics Letters*, 27(1–3):1–3.

were of no interest to him. As will be shown below, this situation changed dramatically during a second respiration interval. The value of these molecules simply lay in the fact that he could obtain control marker location.

> The two dyes need to be a few nanometers apart. The easiest way was to do it with DNA. You can synthesize DNA, you can control the length, you can put a three or five dye anywhere you want, and so on and so forth. Once we got to that conclusion, we looked at who in Berkeley might be able to help us, and we found a colleague of mine, Paul Selvin, who was actually doing first measurements on that sample. So he knew how to prepare the samples, and we started to talk and meet and he said, "look, if you can measure FRET on a single molecule, this is much more than doing spectroscopy. That would be really very useful for biology because you will be able to look at changes. So you can go basically from static structure to dynamic structure. So we knew what we wanted to do, and we knew why we wanted to do it, and we had a really nice goal."[18]

Weiss's shift in this direction is anchored in the new metrological horizons stemming from nanoscale research instruments and the kind of objects they can investigate and observe. The penetration of biosubstances into Weiss's world was surely associated with the massive growth of molecular biology and genetics investigations, in turn associated with the International Human Genome Project (see Chapter 3). Consciousness of these openings was a context for reflection on the limits and potentialities of his past combinatorials, as well as on future research projects in the light of available new combinatorials; including investigation on molecular dynamics versus static. This configuration is suitably seen in terms of the funneling of numerous combinatorials in order to form a well-defined research problem and to mobilize appropriate investigatory resources. Weiss's future research would thus constitute a new expression of his epistemic expectations.

5.2.2 From respiration to a second concentration arena

Investigations on biomolecules brought Weiss to understand that the 2- to 8-nm observational window of FRET is too restricted to observe many of these biomolecules' important characteristics, and that the 300 nm of near-field microscopy were too large. What was required was an intermediary observational platform. In order to study multiple sites inside a molecule, it was necessary to mark those sites with fluorescent dyes sensitive to specific frequencies. Each dye responds to a given wavelength; to excite the dyes, Weiss referred to an earlier technology that had long been central to his research: lasers. However multiple lasers had to be employed in order to excite particular dyes. The advantage was higher precision through triangulation. But a large disadvantage resided in the fact that multiple lasers produced image aberrations and hence resulted in unacceptably distorted images. Weiss remained alert to experimental methods that could overcome this problem. By chance, one day he attended a lecture on quantum dots, where the speaker described his success in rendering these more luminescent by coating them. Quantum

[18] Interview of Shimon Weiss (Paris) by authors, 27 January 2008, 22 June 2008, and 15 April 2010.

dots had been readily available for scientific research since 1993.[19] The notion of using QD as molecular markers was a revelation to Weiss because, depending on their size, the dots generate differently colored light in response to a single excitation laser pulse. Nevertheless, there was a very big problem. Quantum dots are hydrophobic, and biomolecules are exclusively hydrophilic. Working with colleagues from different disciplines, Weiss in his turn developed a coating for these semiconductor dots that is hydrophilic. This new observational possibility and method revolutionized research in many areas of biology. The article that brought together this package of combinatorials of Weiss and colleagues in *Science* in 1998 has been cited more than 4000 times.[20] This technique and associated research questions in biology have become a vast global research arena. The technique consists of illuminating two quantum dots of different sizes, which therefore emit different colors of light, located on two separate components of a molecule, using a single laser pulse. It is now a standard and routine technique for biologists studying the structure and dynamics of macromolecules such as DNA and proteins.

The creation and stabilization of this arena (biomaterials such as DNA and proteins, quantum dots, and molecular structure and dynamics) constituted the substance of the second concentration. Weiss himself contributed importantly to this domain for a decade, and continues to do so even now. For example, he worked on questions associated with protein conformational dynamics, using quantum dots and lasers.[21] This included processes of protein folding, which was emerging as a central question in nanoscale biology.[22] Weiss also explored structural elements of DNA polymers.[23] This second concentration documents the centrality of process in the decision-making and action of much NSR. It illustrates the transverse constancy of epistemic expectations, and the significance of combinatorials and recombinations as well as of funneling in the aforementioned processes.

5.2.3 From a second respiration to a shift to biological questions

Beginning in about 2006, little by little, Weiss engaged in a process that drew him into a second respiration interval. This once again entailed a suspension of commitment to a

[19] M. Reed (1993) Quantum dots, *Scientific American*, 268(1): 118–123.

[20] M. Bruchez Jr, M. Moronne, P. Gin, S. Weiss, A.P. Alivisatos (1998) Semiconductor nanocrystals as fluorescent biological labels, *Science*, 281(5385): 2013–2016.

[21] T. Ha, A.Y. Ting, J. Liang, A.A. Deniz, D.S. Chemla, P.G. Schultz, S. Weiss (1999) Temporal fluctuations of fluorescence resonance energy transfer between two dyes conjugated to a single protein, *Chemical Physics*, 247(1): 107–118.

[22] T. Ha, A.Y. Ting, J. Liang, W.B. Caldwell, A.A. Deniz, D.S. Chemla, P.G. Schultz, S. Weiss (1999) Single-molecule fluorescence spectroscopy of enzyme conformational dynamics and cleavage mechanism, *Proceedings of the National Academy of Sciences USA*, 96(3): 893–898; A.A. Deniz, T.A. Laurence, G.S. Beligere, M. Dahan, A.B. Martin, D.S. Chemla, P.E. Dawson, P.G. Schultz, S. Weiss (2000) Single-molecule protein folding: Diffusion fluorescence resonance energy transfer studies of the denaturation of chymotrypsin inhibitor, 2. *Proceedings of the National Academy of Sciences USA*, 97(10): 5179–5184.

[23] S. Weiss (2000) Measuring conformational dynamics of biomolecules by single molecule fluorescence spectroscopy, *Natural Structural Biology*, 7(9): 724–729.

given set of combinatorials, which then provided the space to evaluate past achievements and future possibilities. This was demarcated by the introduction of fresh combinatorials that also mobilized selective past combinatorials. While in the past, biomolecules were seen as organic substances with no consideration for their in vivo function, Weiss would henceforth deal with these biological molecules in the more general context of the complex processes of life. As a physicist, Weiss here redefined his research in biology in terms of living organisms, and more particularly in initiation, inhibition, and mutation in a DNA transcription in a living organism.[24] Here again, control constitutes a central feature in Weiss's work and it became another combinatorial that extends those already in place in the projects of the prior decade.

In a third and final concentration period, covering the years from roughly 2008 to the present, Weiss has often used techniques like single-photon excitation to modify DNA transcription, in embryo cells of zebra fish, for example. The objective is the understanding and control of mutation of very precise genes. The new orientation toward life processes and his embarking on research into such questions was linked to an increasing interest in matters of health and therapy. Like the prospect of new combinatorials, the tragedy of cancer thus also became a major factor that triggered and oriented this respiration episode.

> Both looking at imaging in the fish and the mice as a way to diagnose, and also the circuit of a normal cell and the circuit of a cancer cell, to see if we can learn enough about them using those fine tools to tell us where things are and get more insight.[25]

Here is one instance of what became the fresh expression of concentration efforts in an established research arena on biotranscription processes. This is part of a research arena that deals both with fundamental life processes and sometimes also gives rise to information relevant to medical diagnosis and treatment. The research line became a basis for co-operation with other research teams in biophysics.[26] In a wider perspective, Weiss's activities led him to the research arena of experimental evolutionary biology,[27] which is concerned with testing hypotheses and theories of evolution by use of controlled experiments. Evolution can be observed in the laboratory as populations adapt to new environmental conditions. With modern molecular tools, it is possible to pinpoint the mutations that selection acts upon, what brought about the adaptations, and to find out how exactly these mutations work. This description corresponds with a major theme

[24] S. Weiss (2006) Probing dynamic structures and molecular interactions at ultrahigh resolution in-vitro and in live cells using single molecules and single quantum dots. In: *The Ultrastructure of Life: A symposium opening the Centre for Ultrastructural Imaging at King's College London*. London.

[25] Interview of P. Weiss (Paris) by authors, 22 June 2008.

[26] Interview of David Bensimon (LPS-ENS, Paris) by authors, 15 February 2010.

[27] Weiss's work shows his interest in medical diagnoses and treatment, which is only loosely connected with his research on experimental works on transcription mechanisms (see for example: G. Iyer, F. Pinaud, J. Xu, Y. Ebenstein, J. Li, J. Chang, M. Dahan, S. Weiss (2011) Aromatic aldehyde and hydrazine activated peptide coated quantum dots for easy bioconjugation and live cell imaging, *Bioconjugate Chemistry*, 22(6):1006–1011)

in Weiss's research from about 2008 to now. It is evidenced in his publications on transcription processes of initiation and elongation, and on the mechanisms of the assembly of biomolecules.[28]

Although working on biological questions, Weiss continues to be in the physics community. He associates very closely with biophysicists and some biologists as well as with chemists and biochemists, but his affiliation with physics is undiluted. Indeed, the respiration-inspired introduction of combinatorials from other disciplines (in this case biology and materials science) generated flexibility in Weiss's home discipline and thus opened fresh perspectives. The edge of the concentration sphere sometimes lies very close to the periphery of the extension sphere. In Weiss's case, the journey has taken him to the outer extreme of the concentration envelope.

Weiss's growing interest in medical diagnoses and therapy potentially entails contact with a very broad and heterogeneous public, which can extend from clinical medicine to pharmacology, diagnosis equipment, and state regulation. In view of his proximity to the edge of the concentration envelope, it would be interesting to see whether, at some period, and during a subsequent respiration, Weiss crosses from concentration to extension, and if so, whether he, like many others, will subsequently circulate back to the epistemic expectations of the concentration sphere.

5.3 GIFTS OF EXTENSION

We observe that, at least in nanoscale research, some scientists shift from the concentration sphere to the extension sphere, and then move back to concentration. This circulation is accompanied by two moments of respiration: one moment as they quit concentration and a second prior to leaving extension. The content and arrangement of the combinatorials structured by epistemic expectations in the concentration sphere are replaced by an alternative constellation of combinatorials that are broader and more contingent. These alternative combinatorials are shaped by the function-horizons of the extension sphere. The combinatorials of the extension sphere stretch from cognition to law, from pedagogy to manufacturing, and from enterprise to medical diagnoses. They are inclusionary, both because of their scope and the organic and often strongly interactive connections between them. Function-horizon constitutes the true motor of extension. Scientists are in part prompted to transfer to the extension sphere by a wish to express their concentration research findings in a range of materials, and among broader and more heterogeneous environments. Thus, multifarious and diversified expressions of concentration-grounded epistemic expectation results comprise a key motivation for scientists to contribute to the

[28] R.H. Ebright, S. Weiss, A. Chakraborty, D. Wang, Y. Korlann, A. Kapanidis, E. Margeat (2009) Single-molecule analysis of transcription, *Faseb Journal*, 23 Meeting Abstract: 202; N. Gassman, S. Ho, Y. Korlann, J. Chiang, Y. Wu, L.J. Perry, Y. Kim, S. Weiss (2009) In vivo assembly and single-molecule characterization of the transcription machinery from Shewanella oneidensis MR-1, *Protein Expression and Purification*, 65(1): 66–76.

extension sphere. To a certain degree, the products of extension may be viewed as "gifts" offered by concentration efforts to an extra-disciplinary, broader environment.[29]

The episode presented below describes the passage of NSR research originally carried out in the concentration sphere to extension, where the swing respiration interval interlaced epistemic and function combinatorials. The components and processes of the extension sphere are explored in detail on the basis of a case study. A nanometric instrument/product was initially developed inside the concentration sphere in the late 1990s, and then largely restructured and adjusted inside extension in 2002; it is known as Dip-Pen nanolithography.[30] The Dip-Pen nanolithography technology involves a device that deposits nanoscale compounds arranged as arrays. Nanodroplets are ordered such that they individually and collectively constitute objects possessing interesting physical properties that invite exploration; they can also be designed in such a way as to express finely controlled, abundant chemically and geometrically based information.[31]

The Dip-Pen device was initially developed for the fabrication of novel research nano-objects and for their exploration. The idea of Dip-Pen nanolithography evolved from the research team of Chad Mirkin's nanoscale research laboratory at Northwestern University. Mirkin is a graduate of Penn State University and is specialized in nanochemistry. The nano institute he directs was one of the first federally funded US nanoscience and technology laboratory. Today it boasts four laureates of the Feynman Nanotechnology Prize: Mark Ratner (2001), Chad Mirkin (2002), Frazer Stoddard (2007), and Georges Schatz (2008). Mirkin has published over 600 articles in his career so far. He founded the journal, *Small*, which has become a leading periodical in the domain of nanoresearch, and he has been a science adviser to President Obama.

5.3.1 A concentration germinated extension instrument

The Dip-Pen nanolithographic tool/product evolved from two of nanoscale research's key inaugural novel research domains, which suddenly arose during the 1980s and 1990s: self-assembling monolayers and near-field microscopy. Mirkin began his career conducting research on self-assembling monolayers (SAMs), a pioneering and significant material in nanoscale research.[32] As seen in Chapter 1, a self-assembled monolayer is an organized layer of amphiphilic molecules in which one end of the molecule, the "head group," shows a specific, reversible affinity for a substrate. SAMs also have a "tail group," the

[29] In the analytic framework of science proposed by Pierre Bourdieu, passage from the concentration sphere to the extension sphere would be interpreted as the investment of concentration cognitive capital in extension in order to turn a profit either in the form of symbolic or material gain that could subsequently be reinvested back in concentration research (P. Bourdieu, 1975) La Spécificité du champ scientifique; P. Bourdieu (2001) *Science de la Science et Réflexivité*).

[30] A. Marcovich, T. Shinn (2011a) Instrument research, tools, and the knowledge enterprise 1999–2009 Birth and development of Dip-Pen nanolithography, *Science, Technology & Human Values*, 36(6): 864–896.

[31] There exist other nanolithography technologies such as NanoImprint Lithography (NIL) or Magnetolithography (ML). <http://en.wikipedia.org/wiki/Nanolithography> (consulted 7 March 2010).

[32] Interview of Chad Mirkin (Northwestern University, Chicago) by Terry Shinn, 20 January 2008.

functional part of the molecule. Under specific conditions, the deposition and localization of SAMs on a substrate can be precisely controlled.

Mirkin began publishing in the early 1990s, and over the course of the decade, he produced more than 20 articles on the topic of self-assembling monolayers, which comprised his principal research themes. Indeed, the rise of SAMs as a current research-object corresponds with the birth of nanoscale research in the mid 1980s. It was the perpetuation of Mirkin's epistemic expectation work on the control of SAMs, and the growing importance of arrays as a key component of nanoresearch in chemistry, that would indirectly lead him to the idea of using these self-assembling monolayers as a deposition compound—which subsequently opened the path to function-horizons and the extension sphere. Arrays involve the repetitive spacing of discrete elements. An array may be composed of duplicated identical elements, or alternatively of patterns of dimensionally, geometrically, chemically, physically, or biologically differentiated repeated elements and combinations. When repetition attains a certain scale, its geometry takes the form of a grid or a network. Arrays are used in science, for example, as the support for the study of different variables of substances.

The second key combinatorial involved in Mirkin's concentration sphere activities was connected with the instrument revolution that took place in this sphere. This revolution entailed the invention of the scanning tunneling microscope and a sister device, the atomic force microscope (AFM), (see Chapter 1). The AFM can identify the position of single or multiple molecules and determine their physical and chemical compositions. In the language of Mirkin, the AFM can read a molecular message. In addition to its metrological functions, the AFM can also be readily employed to control objects by shifting location and by imposing their size and their shape. In an insightful and inventive combination of the AFM instrument and SAM materials, Mirkin conducted concentration sphere research that inverted the customary AFM detection operations. On the basis of his knowledge of the chemistry and dynamical characteristics of SAMs, he employed his AFM to deposit his substances molecule-by-molecule in a highly controlled manner on SAM-compatible substrates. A second AFM could then be used to observe the outcome in a lateral detection mode. The device, named the Dip-Pen, allowed advanced research on molecular behavior, the spatial features of multi-SAMs interactions, and deposition dynamics. All of this was solidly planted in the concentration landscape. Mirkin's use of the twin combinatorials of SAMs and the AFM underpinned his intense investigation of the physical and chemical properties of SAMs when deposited in various environments and onto different substances. As his work progressed, Mirkin gradually conceived his results as a kind of expression of the historical cultural endeavor to "write" and "read"—in this instance on the nanoscale. In his now famous 1999 *Science* article, "Dip-Pen nanolithography,"[33] which has been cited over 1700 times, Mirkin speaks of a pen (the AFM

[33] R.D. Piner, J. Zhu, F. Xu, S.H. Hong, C.A. Mirkin (1999) "Dip–pen" nanolithography, *Science*, 283(5402): 661–663.

tip), ink (SAMs), and parchment (substrate). This metaphor may in hindsight be viewed as an inspiration that presaged his subsequent move to the extension sphere.

But what precisely is the Dip-Pen? When Mirkin's group began work on the Dip-Pen project, the aim was not to develop a nanolithographic instrument. The topic at hand was more general. Part of it included instrument research. As Mirkin and his group recount in their famous January 1999 *Science* article, the inspiration behind the Dip-Pen was first explicitly the solution to a technical problem that hindered the resolving power of the atomic force microscope. The resolving capacity of the AFM operating in an ambient environment was diminished by the formation of a water droplet between the device's tip and a substrate surface. The problem of the water droplet then soon became transformed into one of the fundamental resources of the Dip-Pen lithography device.

Mirkin noticed that the meniscus[34] acts as a mechanism of capillary transport, by which ink-like compounds could be transported molecule-by-molecule from the tip of the AFM to a substrate, in the same way that ink is transferred from a pen to paper. Through controlled deposition of chemical compounds, such as SAMs, Mirkin could thereby fabricate predesigned arrays that would constitute research-objects. In Figure 5.1, the reader can observe the device's tip, the SAM molecules that slide along the tip to the meniscus (in gray), and that are ultimately deposited on the substrate to form the arrays (here represented as thick strips). An important part of the Dip-Pen is the AFM cantilever and its tip. The cantilever controls the distribution and regularity of the depositing of molecular materials. Mirkin perceived that the matching of compounds and substrate constitutes a crucial factor and is important to the control and stability of the substances used in constructing nano-architectures for purposes of research. At the beginning, Mirkin produced nanostructured arrays with his Dip-Pen that integrated eight parallel writing tips, which he refers to as pens. This limited performance somewhat restricted the research capacity of the device. At this juncture, the Dip-Pen found use exclusively within the concentration sphere, where it participated in a deepening understanding of the chemical and

Figure 5.1 Classic DPN mechanism. Molecular ink diffusing from a nanoscale tip to a surface through a water meniscus. Image from Wikimedia Commons.

[34] A meniscus is here an unintended crescent-shape microscopic droplet.

physical features of SAMs (particularly their dynamics), in extending the research capabilities of the AFM, in exploring SAM/substrate chemistry and physics, and in developing complex arrays for further research.

Mirkin's aforementioned article drew much attention inside the scientific community. But how did Mirkin circulate from the concentration sphere to the extension sphere, and what were the components and processes that constituted the respiration interval that accompanied this move? What were the combinatorials that operated in extension research and, finally, what were the inputs entailed in Mirkin's second respiration interval as he circulated away from extension and again became preoccupied with matters of the concentration sphere?

5.3.2 The "ink-pen and writing" metaphor as respiration toward an extension sphere

In 2001, Mirkin presented his nanotechnology findings and his new device, the Dip-Pen, at the famous annual Gordon Conference—initiated in 1931—an assembly of high-level scientists. The Gordon conferences cover the physical sciences as well as the life sciences.[35] The content of Mirkin's lecture was decidedly in the spirit of the concentration sphere, and was directed to an audience of scientists working in his field. At this conference the well-established venture capitalist Chris Anzalone was present. Anzalone was stimulated by Mirkin's metaphor of "pen, ink, and parchment," its materialization in the Dip-Pen inscription device and what he saw as the manifold possibilities of the latter. He immediately contacted Mirkin. Anzalone spoke with Mirkin enthusiastically regarding the many applications he glimpsed for the Dip-Pen and indicated his wish to establish an enterprise to undertake the research necessary to transform Mirkin's research instrument into a multi-purpose commercial product. Anzalone anticipated important applications for the Dip-Pen in the domain of nanobiology.

At this juncture, Mirkin was solidly located in the concentration sphere defined by epistemic expectations, yet Anzalone's proposal stimulated Mirkin's curiosity as to possibilities available in extension spheres associated with function-horizons. This concatenation of events opened the way to respiration for Mirkin. As indicated above, the metaphor that appeared in the *Science* article was not a project, but a vision. The encounter with Anzalone made Mirkin mindful of the possibility of translating his Dip-Pen research instrument into a device that could extend to a variety of concrete implementations. During the respiration moment, Mirkin was only vaguely conscious of the function exigencies and the many different function-horizons that would be entailed in a fully fledged transfer to the extension sphere. He found himself suspended between the technical and intellectual combinatorials of his past concentration exploration, and a set of fresh, unknown combinatorials associated with new publics and novel problems. Anzalone's

[35] <http://www.frontiersofscience.org/> (consulted 3 January 2011).

suggestion to establish an enterprise for the development, production, and marketing of the Dip-Pen propelled Mirkin into extension and, with reference to Dip-Pen-related investigations, temporarily drew him away from concentration to a measurable degree. As will be seen, Mirkin's passage toward the extension sphere would now require him to address multiple audiences, and most immediately, to re-conceive the Dip-Pen in terms of function as opposed to epistemic expectations. This respiration-grounded reflection would necessarily engage Mirkin in the assembly of many new categories of technical and non-technical combinatorials.

With two other venture capitalists, Mirkin and Anzalone established the firm NanoInk Inc., in 2001.[36] In the extension sphere, Mirkin's first task consisted of transforming his very fickle Dip-Pen research instrument into a reliable device. In doing this, he shifted from epistemic expectations to a preoccupation with technical function. Reliability became his primary referent. Reliability opened the way to his second extension task. The function capabilities of the soon to be manufactured and marketed Dip-Pen nanometric tool also had to address many different economic and social universes.

Mirkin's participation in function aspects of the Dip-Pen resulted in the introduction of four new combinatorials: metallurgy / chemistry, mechanics, computers and biological substances. He initially dealt with failings in the quality of the writing tips of his AFM and with the unsatisfactorily small number of pens that could be mobilized simultaneously. As mentioned earlier, the device initially accommodated a mere eight pens, which was too limited for many applications. Mirkin and his team improved the quality of the tips, investigated difficulties with cantilevers and, above all, multiplied the number of parallel operating pens. Working together with engineers of the NanoInk Company, Mirkin and his colleagues introduced a 55,000-pen tool in 2004. The issue of function was still not fully resolved however. The need to introduce substances with different chemical and physical characteristics, the presence of different substrates, and above all, a variety of requirements in the production of arrays, for example in biology, made the writing and reading process extremely complex. Mirkin became engaged in the utilization of computers in the operation of the Dip-Pen, where the kit made it possible for non-specialists to use the apparatus for the production of highly elaborate scientific and medical nanometric arrays.

Most notably, Mirkin's efforts in the extension sphere led him to develop expertise in nanobiology and the life sciences, which quickly emerged as a major combinatorial. Detection and diagnosis in the health sciences and much DNA and protein basic nanobiology research employed arrays, either as a basis for experimentation or for clinical work. Since one of the principal fields of application is array production, logically, Mirkin was increasingly drawn into activities associated with biological matters. As will be seen, this addition of the biology combinatorial would soon figure importantly in a second respiration interval and in Mirkin's circulation away from the extension sphere and back into the concentration sphere.

[36] <http: / / www.nanoink.net/ > (consulted 2 November 2012).

Indirectly, Mirkin would encounter audiences from many horizons, since arrays are employed for detection and investigation in numerous scientific and industrial fields. The extent of the universes involved with the Dip-Pen is reflected in the departments established in NanoInk Inc. The company contains five divisions: the NanoFabrication System Division, the NanoBio Discovery Division, the NanoStem Cell Division, the NanoGuardian Division, and the NanoProfessor Division.

The first unit to be created was the NanoFabrication System Division. Its initial task was the transformation of instrumentation research into a commercially viable tool for the production of nanometric arrays. This division builds standardized Dip-Pen tools in the form of kits. Two of the five divisions deal specifically with biological materials: the NanobioDiscovery Division uses the Dip-Pen to manufacture arrays specifically adapted to biological research, which includes, in combination with proprietary detection systems, the production of uniquely sensitive slides for protein discovery and identification.[37] The second biological division, the NanoStem Cell Division, is once again grounded on the construction of chemically and geometrically highly complex nanometric arrays. In this division, the Dip-Pen technology is used to render precise nanopatterns capable of producing a homogeneous population of differentiated adult cells. Adherence, growth, and differentiation of adult cells are crucially dependent upon both the composition of the chemical cue and the pattern of the cues on the biochip. Examples of stem-cell growth given by NanoInk include brain, fat, bone, and cartilage cells.[38]

Unlike the previously described division, the activities of the NanoGuardian Division are unrelated to research, but are instead concerned with the application of NanoInk technology to the output products of mass-scale industry. This division uses Dip-Pen based overt and covert inscription technology for traceability of consumables, of the manufacturing trajectory of products such as medication, and to protect companies against counterfeiting. For example, in the area of pharmaceuticals, it inscribes at the nanoscale, information concerning the production and the date and site of manufacture, identification of the specific product, and expiration date for the dose of medication.[39] The NanoProfessor Division is involved in training. It provides Dip-Pen kits for secondary and university education in the fundamental principles of nanotechnology.[40] The product also familiarizes students with techniques for the fabrication of nano-arrays, which are seen by some as constituting a massive employment market.

By 2004–5, NanoInk Inc. had begun to stabilize its fabrication and commercial activities, and was setting out on programs of further developments. During the early days, a measure of Mirkin's considerable energy was linked to the evolution of the Dip-Pen

[37] <http://www.nanoink.net/d/appnote_JustAddDNA.pdf> (accessed 11 March 2009).

[38] <http://www.nanoink.net/NanoStem_about.htm> (20 September 2009).

[39] <http://www.nanoink.net/NanoGuardian_about.htm> (22 September 2009).

[40] Telephone Interview of Tom Levesque by authors, 5 March 2009. At the time of the interview, Tom Levesque was director of the NanoFabrication Division.

tool inside the firm and beyond. In these practices, the scientist met with many function-horizons. It would be inaccurate to suggest that Mirkin was directly involved in the activities of all five divisions of NanoInk Inc.; it is nevertheless the case that he engaged in selective technical projects for the company and that, when appropriate and necessary, his commitment to function-horizon issues extended far beyond technology. Mirkin's subsequent status as a science adviser to President Obama is emblematic of his engagement in function-horizons and the extension sphere.

5.3.3 Side-stepping back to concentration—toward nanobiology

In part inspired by his experiences in the extension sphere and particularly in areas of nanobiology, Mirkin travelled back to the concentration sphere, where he was once again preoccupied by epistemic expectations. The switch from extension back to concentration was accompanied by a brief respiration moment. It appears that his re-orientation toward the concentration sphere was a combination of a wish to focus on new scientific and technological combinatorials and a desire to return to older problematics and materials.

Mirkin's new efforts in the concentration sphere adapted the nanobiology-related combinatorial that had emerged during extension research to narrowly epistemic objectives. For example, he generated fresh combinatorials that emphasized investigations of DNA and proteins,[41] and moreover he linked biological issues to, for him, new non-organic substances such as nanorods and nanotubes.[42] These represented still more combinatorials. Finally, Mirkin adopted a supplementary combinatorial in the form of simulation and methodology as a means of powerfully converging his mental and material resources onto a restricted range of epistemic issues[43] (see Chapter 2).

Mirkin's work draws attention to the possible complementary interaction and reciprocal nourishing of the concentration and extension spheres in NSR. The significance of respiration operating between and within the two spheres lies in the fact that it offers an interval for assessing the past and looking to the future in terms of antecedent combinatorials and the capacity to mobilize new ones. In respiration, scientists find the space to reflect on the orientation and content of their curiosity and to decide whether this curiosity finds a hospitable terrain inside concentration or inside extension. Alternatively, some scientists, as we have seen in the case of Mirkin, discover that both spheres offer hospitality. In effect, respiration is a moment where practitioners find a measure of freedom to redirect their path or to affirm their present route. It is a particularly dynamic phase of research activity.

[41] S.W. Lee, B.K. Oh, R.G. Sanedrin, K. Salaita, T. Fujigaya, C.A. Mirkin (2006) Biologically active protein nanoarrays generated using parallel dip-pen nanolithography, *Advanced Materials*, 18(9): 1133.

[42] K.B. Lee, S. Park, C.A. Mirkin (2004) Multicomponent magnetic nanorods for biomolecular separations, *Angewandte Chemie—International Edition*, 43(23): 3048–3050.

[43] S. Zou, D. Maspoch, Y. Wang, C.A. Mirkin, G.C. Schatz (2007) Rings of single-walled carbon nanotubes: Molecular-template directed assembly and Monte Carlo modelling, *Nano Letters*, 7(2): 276–280.

CONCLUSION

The trajectory and temporality of the research projects and orientations of many NSR practitioners are punctuated by intervals of interruption. Such punctuations—respiration—reflect important cognitive features characteristic of NSR. As developed in this chapter, respiration consists of brief suspensions in research activity when scientists examine the future potential of current lines of research in terms of intellectual, material, and methodological resources, before either assembling additional capacity to carry on, or assembling alternative and broader material and cognitive resources judged to be useful for turning to a new direction.

We have suggested that the occurrence of respiration in NSR is largely a product of the centrality of an unending stock of combinatorials in the field. As indicated in Chapter 1, the stuff of combinatorials comprises both the instrumentation and the materials mainstay of NSR since the genesis of the domain; and combinatorials similarly figure centrally in most individual research projects. It is the volume and availability of combinatorials that both impose respirations on NSR scientists, and are viewed as a reservoir enabling prolific new conditions of possibility.

In NSR, the practice of respiration has given rise to several geometries of research structure and strategy. Respiration functions inside what we have called the concentration sphere, where epistemological advance comprises the principal expectation. Here, respiration represents moments during research processes when practitioners reflect on the potential of their combinatorial resources, in terms of continuing a project or shifting to a new set of questions. In the one instance, scientists persist in their former arena, and in the other instance, while continuing inside the concentration sphere, they transfer to an alternative arena.

Beyond the concentration sphere, some NSR practitioners engage trajectories that take them into the extension sphere. The scope of combinatorials inside this sphere is both different and broader than concentration. As shown in this chapter, the research framework of extension is the function-horizon. Research is structured in terms of attaining practical outputs as opposed to epistemological ends. In this environment, combinatorials often include economic, organizational, bureaucratic, and ideological components, alongside more technical ones. In most instances, circulation of scientists begins with concentration, moves to extension, and then later engages concentration.

6

Which Disciplinarity for Nanoscale Research?

There is considerable debate between those who believe that contemporary scientific research is characterized by "interdisciplinarity" and those claiming that the organization and structures of scientific practice instead remain primarily disciplinary in orientation. In this chapter, we explore where NSR stands in this controversy. Interdisciplinarity insists on the absence of boundaries between bodies of learning, on entirely fluid circulation of practitioners, and on "integrated" forms of cognition.[1] In contrast, stated simply, disciplinarity suggests the maintenance of disciplinary referents that frame learning, questions, methods, instrumentation, and evaluative criteria. Here, boundaries function to set off disciplines from one another. The disciplinary referent and boundary define and regulate disciplinary endeavor.[2]

[1] The theme of interdisciplinarity versus disciplinarity has received a great deal of attention among scholars and science policy-makers. Among others we can cite: R. Frodeman, J. Thompson-Klein, C. Mitchan (2010) *The Oxford Handbook of Interdisciplinarity.* Oxford: Oxford University Press; M. Gibbons, C. Limoges, H. Nowotny et al. (1994) *The New Production of Knowledge: The dynamics of science and research in contemporary societies*. London: Sage; M. Lamont, G. Mallard, J. Guetzkow (2006) Beyond blind faith: Overcoming the obstacles to interdisciplinary evaluation, *Research Evaluation*, 15(1): 43–55; L. Leydesdorf, P. Zhou (2007) Nanotechnology as a field of science: Its delineation in terms of journals and patents, *Scientometrics*, 70(3): 693–713; M. Meyer, O. Persson (1998) Nanotechnology—interdisciplinarity, patterns of collaboration and differences in application, *Scientometrics*, 42(2): 195–205; H. Nowotny (with P. Scott, M. Gibbons) (2001) *Re-Thinking Science. Knowledge and the public in an age of uncertainty*. Oxford: Polity; I. Rafols (2007) Strategies for knowledge acquisition in bio-nanotechnology: Why are interdisciplinary practices less widespread than expected? *Innovation: The European journal of social science research*, 20(4): 395–412; M.C. Roco (2003) Nanotechnology: Convergence with modern biology and medicine, *Current Opinion in Biotechnology*,14(3): 337–346; M.C. Roco (2004) Nanoscale science and engineering: Unifying and transforming tools, *AIChEJ.*, 50(5): 890–897; M.C. Roco, W.S. Bainbridge (2003) *Converging Technologies for Improving Human Performance: Nanotechnology, biotechnology information technology and cognitive science.* Dordrecht: Kluwer Academic Publishers; J. Thompson-Klein (1991) *Interdisciplinarity. History, theory and practice.* Detroit, MI: Wayne State University Press; L. Thompson Klein (1996) *Crossing Boundaries: Knowledge, disciplinarities, and interdisciplinarities*. Charlottesville, VA/London: University Press of Virginia; J. Thompson-Klein (2005) *Humanities, Culture and Interdisciplinarity. The changing American academia*. New York, NY: State University of New York.

[2] P. Bourdieu (1984) *Homo Academicus*. Paris: Minuit; P. Bourdieu (2001) *Science de la science et reflexivite*. Paris: Raisons d'Agir; G. Lemaine, M. Mulkay, P. Weingart (1976) *Perspectives on the Emergence of Scientific Disciplines*. Paris: Mouton; S. Turner (2000) What are disciplines? And how is interdisciplinarity different?

Which of these models most adequately fits the activities of NSR, or must one instead seek out an alternative model? In the first section of this chapter, six parameters will be deployed to describe and analyze our empirical observations of NSR activities. These parameters include: (1) the structure and operation of what we term the "referent" that likens the entity habitually labeled "discipline"; (2) how the volume of combinatorials in NSR affect the activities of referents; (3) the importance and function of "projects" in NSR, and their impact on the workings of a disciplinary referent; (4) the introduction of a "borderland" in parallel with disciplinary boundaries, and how it operates cognitively; (5) how NSR practitioners circulate about inside their home disciplinary referent when torn between project/borderland imperatives and more constant referent-based imperatives; (6) finally, "displacement" introduces the parameter of "temporalities," alien to discipline descriptions of research work. We will suggest that what we here term "the new disciplinarity," which emerges from our NSR exploration, may indeed even today occur in other fields of investigation, where the orthodoxy of aspiring interdisciplinarity and insistence of classical disciplinarity no longer offer adequate description. In the second section of the chapter we set forth four typical trajectories of NSR practitioners: (1) beckoning from indoors; (2) extending the repertory; (3) common territory for answering homeland questions; and (4) successive projects, which intertwine referent, combinatorials, projects, borderlands, displacement, and temporalities.

6.1 THE "NEW DISCIPLINARITY" IN ACTION

6.1.1 The disciplinary referent

Disciplines constitute the stable referent of practitioners, providing the intellectual and material resources, language, and universe of questions. In the specific case of the "new disciplinarity," it is from this territory that scientists may engage in projects that provisionally associate them with other disciplines. As will be shown in this chapter, referent looms large in the practices of NSR researchers. However, unlike other areas of science, NSR exhibits an ensemble of practices that are not adequately described in the traditional framework of discipline. The discipline constitutes a practitioner's home referent: it is the discipline from which he issues; back to which that he returns; and it is the discipline that provides a practitioner with a set of coordinates that determine his course and his identity. Identity is an assertive element for the practitioners of a discipline. It is linked to the concept of affiliation, developed through years of immersion in a cognitive corpus having specific concepts, a particular vocabulary, being grounded in a family of phenomena,

In: P. Weingart, N. Sterh (eds.) *Practicing Interdisciplinarity*. Toronto: University of Toronto Press, pp. 46–65; A. Abbott (1995) Things of Boundaries, *Social Research*, Vol. 62(4), pp. 857–882.

A. Abbott (2001) *Chaos of Disciplines*. Chicago and London: The University of Chicago Press; J.A. Jacobs (2013) *In Defense of Disciplines. Interdisciplinarity and Specialization in the Research University*. Chicago: The University of Chicago Press,.

materials, instruments, theories—all reinforced by "habitus" and the professional associations of practitioners. The discipline provides broad integrated learning in conjunction with in-depth specialization.[3] Cultural continuity is associated with socialization acquired through writing dissertations, teaching, research production, productivity in the form of publications, and recognition. These elements are all part of the process which formats a disciplinary trajectory and associated expectations.

Disciplinary identity of practitioners often appears to be indelible: when working with practitioners from other fields, a scientist retains his intellectual and institutional coloration, his identity. Each party to the collaboration contributes from their own particular vantage point. The practitioners stay in their respective space and exhibit the characteristics, ethnographic markers, associated with that disciplinary space.

Of major significance, the language of a discipline's practitioners entails a specific, constrained vocabulary, which expresses the authorized corpus of the discipline, and at the same time often renders unintelligible the substance of other disciplines.[4] Here a discipline's language is the vehicle of intra-disciplinary communication and a solid obstacle to communication and intelligibility with other domains. Even if a scientist tries to understand the language of the other disciplines, he keeps his "mother tongue" and most of the time has to "transcribe" his ideas, questions and knowledge into the other discipline's language. The discipline thus little by little comprises a person's intellectual and social territory. The practitioner learns to know its contours; grows comfortable with them; they provide the coordinates for the scientist's situatedness.

The connection between discipline and individual practitioner action necessarily lies foremost in the capacity to ask a truly penetrating and original research question. Practitioners tell us that it is impossible for them to ask a good question outside of their referent discipline, even if they have been considerably involved in the work of other domains.[5] In domains other than their own, they only assimilate specifically what they need in order to solve the immediate problem at hand. The new knowledge is strictly instrumental and is not enriched or contextualized by a broader science culture. Stated differently, in the absence of fluency in the language of another discipline, practitioners indeed lack the necessary resources for full proficiency. The possession of an instrumental, fragmentary vocabulary proves to be insufficient for introducing truly good questions. Limits of the human mind and limits in time constitute a material barrier that constrains. Put simply, not everything is possible!

Our above description of referent largely corresponds with the treatment of discipline by many historians and sociologists of science, who focus on the organization of

[3] R. Whitley (2000) *The Intellectual and Social Organization of the Sciences*. Oxford: Oxford University Press.

[4] P. Galison (1997) *Image of Logic. A material culture of microphysics*. Chicago, IL: University of Chicago Press.

[5] Interview of Shimon Weiss, professor at UCLA and the Maison des Sciences de l'Homme Paris, by authors, June 2008.

nineteenth and early twentieth-century academic scholarship.[6] A foundational characteristic of this expression of discipline is that it was often defensive, being inward looking and self-protective. From this posture, disciplines were thus closed. This frequently implied an inclination toward the static. This profile is what we refer to as "traditional disciplinarity." Traditional disciplinarity differs from the new disciplinarity, which entails a balance between stability and mobility, and incorporates structures that authorize communication and collaboration with other disciplines. The remainder of our chapter deals with the introduction of additional dimensions that strongly modify the place of referent without however eliminating it. As will be seen below, this constitutes the "new disciplinarity," and this new disciplinarity is a defining feature of nanoscale research. In NSR, we observe that an outward-looking gaze is a necessity, which draws attention to the role played by combinatorials.

6.1.2 Combinatorials

This book has throughout documented the centrality of combinatorials in NSR. Combinatorials, as will be recalled, are the association of elements (instruments, materials, concepts, and people) whose novel integration constitutes often powerful resources for extending problem solving. Stated differently, a threshold of recently introduced instruments, materials, data, ideas, and applications, both in quantity and diversity, is a positive environment for the emergence of combinatorials. They can be thought of as a matrix of possible permutations and associations. Combinatorials prove so very powerful in the originality and extension of research, because of the possibility of fresh associations and integrations of devices they offer. They prove innovative in the stimulation of questions, offering realistic paths of problem solving, and in introducing new links between otherwise distant fields and communities. Thus, by presenting possibilities for fresh combinations of material and cognitive permutations, combinatorials can generate a new repertory, vocabulary, and vision.

Among the most important elements contributing to the expansion of combinatorials in research during recent decades, one must point in particular to the important contribution of computer simulation.[7] Here the question is not of using, in a routine fashion, an instrument that was first intended for extremely well-defined purposes (the Manhattan Project, flight engineering and training, gunnery training and, to a great degree, graphics

[6] J. Ben-David (1971) *The Scientist's Role in Society, A comparative study*. Englewood Cliffs, NJ: Prentice-Hall; T. Lenoir (1997) *Instituting Science. The cultural Production of Scientific Disciplines*. Stanford, CA: Stanford University Press; R. Whitley (2000) *The Intellectual and Social Organization of the Sciences*. Oxford: Oxford University Press.

[7] J. Kueppers, J. Lehnard, T. Shinn (eds.) (2006) *Simulation: Pragmatic construction of reality*. Dortrecht: Springer; P. Humphrey, C. Imbert (eds.) (2012) *Models, Simulations and Representations*. London: Routledge; J.M. Duràn, E. Arnold (eds.) (2013) *Computer Simulations and the Changing Face of Scientific Experimentation*. New Castle: Cambridge Scholars Publishing.

for the motion picture industry).[8] Instead, it is to employ it in close combination and collaboration with a vast range of other research devices and methodologies. This instrument is mobilized by scientists whose skill, knowledge, and even expectations are necessary for the research project to which it is added, and proves opportune, and often even essential. Simulation figures perhaps more regularly in combinatorials than any other single component. In nanoscale research, it is combined with the scanning tunneling microscope, the atomic force microscope, the near-field optical microscope, laser detection of markers, and analytic mathematics. It is increasingly used in the study of biological nanomaterials, where simulation permits scientists to select particular codes for experimentation or suggest promising conceptual paths (see Chapter 3).[9] Simulation is today so ubiquitous in scientific research that it too often goes almost unmentioned. Yet because of its powers of representation and analysis, it stands foremost in many combinatorials. It is safe to say that simulation, like few other methods/devices, contributes massively to the symphony of combinatorials.

The ubiquity of combinatorials in NSR opens the way to a world of expanding complexity. New categories of objects are observed. New perspectives develop with relation to new objects, as well as to historical ones. Complex interactions between specialties and disciplines arise. Recalling that NSR practitioners sustain their privileged links with their disciplinary referent, how do they manage the complexity of their subsequent ensuant combinatorials? What new disciplinary structures and interactions are developed? In the following subsections, we document the place of projects in NSR and the role played by what we term "borderland" and "displacement."

6.1.3 Projects and borderlands

Projects are a crystallization of questions that are asked in the framework of one discipline, but which require resources belonging to other disciplines in order to solve the problem. Otherwise stated, they may be regarded as the domain for the upgrading of questions. This is all part and parcel of the combinatorials and complexity issues that are so characteristic of NSR. Projects and combinatorials then contribute to open a path of communication between disciplines where exchange of ideas, questions, and skills can lead to productive problematics, yet without dissolving the authenticity of the discipline. The territory thus opened is the territory where encounters and communalities can crystallize ideas and questions, and where new disciplinarity projects

[8] T. Shinn (2006) When is simulation a research technology? Practice, markets and lingua franca. In: J. Kueppers, J. Lehnard, T. Shinn (eds.) (2006) *Simulation: Pragmatic construction of reality.* Dortrecht: Springer, pp. 187–205.

[9] A. Marcovich, T. Shinn (2010) Socio/intellectual patterns in nanoscale research. Feynman Nanotechnology Prize laureates, 1993–2007, *Social Science Information*, 49(4): 1–24; A. Marcovich, T. Shinn (2013) Computer simulation and the growth of nanoscale research in biology. In: Duràn, J.M. and E. Arnold (eds.) (2013) *Computer Simulations and the Changing Face of Scientific Experimentation*, pp. 145–170.

can be elaborated. Projects emerge in the effervescence of communications and of shared instruments, more broadly speaking, elements arising in the disciplinary research arena.[10]

How many times during our study have we observed questions that were raised by scientists in the nanoscale perspective which mobilized material and intellectual cooperation with colleagues in neighboring domains? The crystallization of new efforts was defined in terms of cooperation and not in the framework of the abandonment of a home discipline. Through projects, cognitive complementarity develops, as opposed to interdisciplinarity with its attendant supposed integration of knowledge.

How do cooperative projects among practitioners from different disciplines develop and unfold? Through what mechanism is disciplinary adherence sustained, which also permits multi-disciplinary communication? In effect, how is it possible both to remain in a discipline and to generate meaningful interactions with individuals from other disciplines? We introduce the concept of borderland to account for the compatibility of disciplinary maintenance and cross-disciplinary action. The existence and operation of borderland distinguishes the new disciplinarity from traditional disciplinarity, which obeys a logic of powerfully exclusionary demarcated boundaries.

The borderland stands at the periphery of a discipline. It is not to be confused with a "boundary." A boundary is a blockage device which prevents exit and entry. A boundary line is not always agreed or exactingly precise. There frequently exists a small measure of ambiguity in the precise trajectory of a border. This is particularly problematic for a person in rough territory. It is this situation that gives rise to what is termed a "borderland," which is an indefinite, fuzzy, narrow swath of terrain contiguous with the established borders of two recognized entities—in our case, scientific disciplines.[11]

In NSR, scientists desiring extra-territorial contact (extra-disciplinary contact), each one standing on that patch of the borderland contiguous with his territory and addressing a practitioner symmetrically located on his own borderland adjacent to his particular territory, initiate and pursue communication from this strategic site. The existence of borderlands at the periphery of disciplines constitutes the space that permits scientists to approach one another freed from the necessity of confronting or breeching the sturdy walls that continue to distinguish disciplines. Borderlands indirectly allow practitioners to speak from the safety of their disciplinary homelands. It is from here that they engage in activities of extra-disciplinary projects. The two scientists are, for a particular project and for a short period, sufficiently proximal as to be capable of selective, useful exchange and interaction.

Trans-boundary exchange may be represented in a variety of ways: it may be seen as practitioners of separate disciplines, each standing in his discipline's own area of their

[10] W. Espeland, M. Stevens (1998) Commensuration as a social process, *Annual Review of Sociology*, 24: 313–343.

[11] R.E. Kohler (2002) *Landscapes and Labscapes: Exploring the lab—field border in biology*. Chicago, IL/London: University of Chicago Press.

own borderland, and each versed in a particular language, shouting back and forth to each other across the dividing border wall. Alternatively, borrowing the image of "handshaking" advanced by the philosopher of science, Eric Winsberg, scientists working with quite different problems each in their specific frame, may reach over the dividing wall and temporarily join hands (for the duration of the project) in seeking a common denominator that can enhance intelligibility for all.[12]

6.1.4 Displacement and temporality

Displacement refers to a selective intermittent movement of a scientist into the borderland of his discipline and a subsequent retreat when he circulates back to his disciplinary nucleus. The discipline constitutes the referent and acts as a constant attractive gravitation-like force on the practitioner. While powerful, the force is nevertheless not so strong that it prevents the individual from provisionally distancing himself from the center and instead turning his gaze toward the borderland and thus to the periphery of his home discipline. Displacement is the rule in NSR, and this is what guarantees the referent and hence sustains authenticity. Temporality describes a relationship between discipline and projects, and it is inextricably connected with practitioner displacement. There are two major categories of temporality: long term and short term. Long-term temporality applies to the referent discipline where individuals spend most of their time. Short-term temporality is connected with projects, which tend to be short-lived events when compared with the time devoted to the home discipline-based endeavors. Scientists cycle back and forth between the heartland of their discipline and the periphery and project. In NSR, this is not experienced as alienating because the scientist sees the project as a site for acquiring and for expressing combinatorials.

The status of the two forms of temporality is equal. Practitioners understand that both are essential to research, since it is desirable to participate in both the work of the disciplinary nucleus and project operations. The two temporalities are thus complementary and do not constitute a source of tension. Whatever the form, temporality is an integral component of displacement, which allows individuals full and legitimate participation in peripheral borderland territories, while retaining the discipline as the foundational referent.[13]

[12] E. Winsberg (2006) Handshaking your way to the top: Simulation at the nanoscale, *Sociology of Sciences Yearbook*, 25(Part 3): 139–151.

[13] In an earlier article, "Where is disciplinarity going?," we reflected on the dynamics involved in the birth and development of new disciplines. This topic is a thorny question and did not directly concern the issues discussed here. We intend to return to them in the near future. The recent book by Kostas Gavroglou and Ana Simões, *Neither Physics nor Chemistry,* raises the interesting concept of the introduction of a new discipline in terms of an in-between discipline, and this orientation certainly warrants additional investigations. A. Marcovich, T. Shinn (2011). Where is disciplinarity going? Meeting on the borderland, *Social Science Information*, 50(3–4): 582–606; K. Gavroglou, A. Simões (2012) *Neither Physics Nor Chemistry.* Cambridge, MA: The MIT Press.

6.2 COGNITIVE TRAJECTORIES

Our interviews of 49 nanoscale scientists reveal the prevalence of four categories of trajectory; three of these explicitly fit inside the framework of the new disciplinarity. Only one category, the first of these, reflects many of the cognitive characteristics associated with what we above refer to as traditional disciplinarity, where practitioners reside deep in the homeland of their specialty or discipline, and only rarely if ever shift toward the periphery and the borderland. We label this trajectory "beckoning from in-doors." In a second NSR trajectory, "extending the repertory," practitioners based in their specific field travel to the borderland and to a limited extent cooperate with scientists from different domains as they search for materials that will enable them to broaden the problematic that runs as a cognitive thread throughout their career. In this process, selective elements are drawn from outside the homeland that do not require assimilation of knowledge extraneous to the hub discipline. In the third configuration, "common territory for answering homeland questions," practitioners from different specialties circulate to their particular borderland in search of questions, techniques, and materials that will enable them to address respective referent homelands issues from a new and broader perspective. They circulate freely through their disciplinary hub to borderlands where collaborations give a new breath to established questions. Finally, the fourth trajectory, "successive projects," illustrates a long-term, balanced dynamics between homeland and projects. In the example presented below, a nanoscientist repeatedly visits borderlands. He sustains his initial cognitive coordinates and at the same time draws significantly from selective aspects of the learning from other specialists, who enrich and to an extent redirect his path. These successive borderland projects take on a logic of their own.

6.2.1 Beckoning from in-doors

In this trajectory, the referent constitutes the primary determining orientation. The concepts of borderlands and project as understood in the new disciplinarity are absent. This implies that displacement is not an issue, and the measure of temporality is only that of the entire career. In many important respects, this profile is reminiscent of traditional disciplinarity.

Beckoning from in-doors researchers may be seen as gazing outward from the window of their discipline and beckoning with a gesture of their hand to outsiders to come and take an interest in their work. The trajectory of Fraser Stoddart exemplifies this beckoning from in-doors category.

Stoddart was born in Scotland and is a chemist, presently working in the United States at Northwestern University, in the nanoscale research laboratory directed by Chad Mirkin (see Chapter 5). Stoddart is the laureate of the 2007 Feynman Nanotechnology Prize. He has developed a technique for the synthesis of mechanically interlocked molecular architecture, such as molecular Borromean rings, catenanes and rotaxanes, utilizing

Figure 6.1 Schematic of a molecular Borromean ring. Image from Wikimedia Commons.

molecular recognition and molecular self-assembly processes.[14] In the following figures, one can see two simulation designed Borromean rings. In both, three rings are interlocked. The simplicity of Figure 6.1 permits an appreciation of the underlying logic and complexity of Figure 6.2. In the following detailed description of Stoddart's work, the reader will notice that there is no reference to other questions or even substances other than his. Research carried out in alternative disciplines, and even research in neighboring areas of chemistry, receive no attention. The gaze is consistently, directly in-doors.

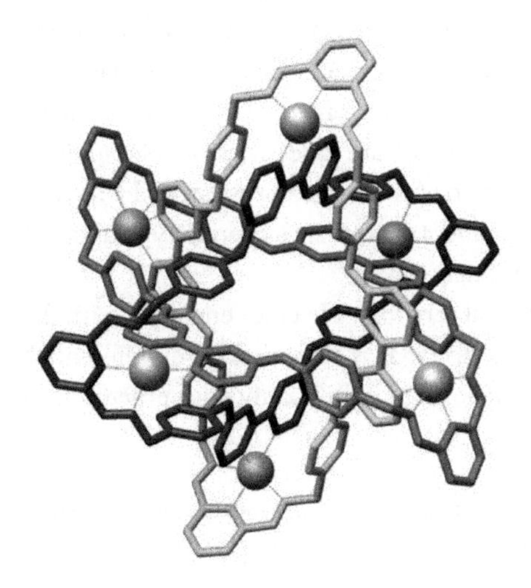

Figure 6.2 Molecular Borromean ring. Image from Wikimedia Commons.

[14] K.S. Chichak, S.J. Cantrill, A.R. Pease, S.H. Chiu, G.W.V. Cave, J.L. Atwood , J.F. Stoddart (2004) Molecular Borromean rings, *Science*, 304(5675): 1308–1312.

Stoddart has demonstrated that these architectures can be employed as molecular switches and as motor-molecules. His group has even applied these structures in the fabrication of nanoelectronic devices and nanoelectromechanical systems (NEMS).[15] He has authored over a thousand articles, among which "Interlocked and intertwined structures and superstructures,"[16] published in 1995, has been cited 1171 times.

> We discovered by a process of trial and error and research of self-assembly and molecular recognition how to make interlocked rings; if you have two interlocked rings they're called catenanes, if you lock a ring around a dumbbell component then you make a rotaxane. That progress was made during the 1980s. By the end of the 1980s and the beginning of the 1990s, we were able to show that we could put together many rings one after the other in a chain, and that we could put many rings on a dumbbell. To be specific, we made one compound that entailed five rings as enfolded Borromean rings. This first generation of compounds were degenerate. The rings moved discretely one with respect to another and we could follow this process by spectroscopy, nuclear magnetic resonance spectroscopy, but what we needed to do and what we addressed through the period of the mid-1990s, until I came to UCLA in 1997, was to introduce the potential of switching into the molecules.[17]

During roughly the first decade of his career, Stoddart, like so many other chemists of this period, was actively engaged in research related to the elaboration of chemical synthesis, which covered a range of substances and issues. For roughly the two decades after 1990, an examination of Stoddart's remarkable publication list reveals that his attention has focused on a well-defined set of nanochemistry questions and problems. Omnipresent considerations have included items like reversible bindings, recognition, self-assembly, dynamic processes, and electron activation (see Chapters 1 and 4). One of Stoddart's most cited articles, "Electronically configurable molecular-based logic gates,"[18] incorporates and discusses most of these elements. A kind of "mechanical" vision of nanoscale chemistry is nourished by these different elements. The mechanical vision is perhaps best expressed and crystallized in the notions of switching and control. The terms "switch," "recognition," "meccano," "Rotaxane," and "catenane" are categories of nano-objects that represent the universal building blocks of Stoddart's chemistry system and which he refers to as "lego" constructions (see Chapters 3 and 4). Stoddart's chemistry consists of interlocking these categories of objects in a designed way.

Stoddart dominates this cognitive and technical territory. He deals endlessly with the same objects and in the same perspective. The research is remarkably self-referential. There is no evidence to suggest that he has collaborated significantly with practitioners

[15] Nanoelectromechanical systems (NEMS) are devices integrating electrical and mechanical functionality on the nanoscale. <http://en.wikipedia.org/wiki/Nanoelectromechanical_system> (consulted 24 September 2011).

[16] D.B. Amabilino, J.F. Stoddart (1995) Interlocked and intertwined structures and superstructures, *Chemical Reviews*, 95(8): 2725–2828.

[17] Interview of Stoddart, James Frazer (UCLA) by Terry Shinn, 29 January 2008.

[18] C.P. Collier, E.W. Wong, M. Belohradsky, F.M. Raymo, J.F. Stoddart, P.J. Kuekes, R.S. Williams, J.R. Heath (1999) Electronically configurable molecular-based logic gates, *Science*, 285(5426): 391–394.

from other disciplines such as biology or physics, or even specialists from other areas of chemistry. This is not to say, however, that he has not looked beyond his specialty—that he has not looked outside the window of his particular domain. On the contrary, he often looks outside the window, but when doing so, he beckons to outsiders in order to draw attention to the significance of his nano work and to make it understandable. Stoddart stands deep in the heartland of his own discipline and rarely, if ever, even glances toward the periphery, borderlands, or projects. Although he is certainly an accomplished and renowned nanoscientist, who could better represent the stability of traditional disciplinarity?

6.2.2 Extending the repertory

One can observe, for a certain category of nanoresearchers, that a set of questions structures a great part of their work. In the course of this pursuit, they intermittently move to the borderland of their discipline, as they seek cognitive resources intended to extend their home repertory. In so doing they do not attempt to become learned in an outside domain. Contacts are largely short term.

During the opening fifteen years of his complex career, James Gimzewski operated from deep inside his referent physics discipline. He explored the properties of an extensive range of nano-objects, and at the same time conducted research on nano-instruments and associated physics principles (see Chapter 2). He subsequently utilized his metrology knowledge to develop novel instruments in the landscape of narrowly set physics domains. This perspective constituted his initial research repertory. Here Gimzewski stood deep inside his discipline. He subsequently, intermittently moved between the periphery and the heart of nanophysics as he became engaged in the exploration of a variety of materials and questions located in chemistry and biology. Through this opening, he extended his earlier established repertory. Joint research with practitioners from non-physics fields was required, and took the form of projects. Each scientist operated from inside his particular disciplinary frame. Projects were marked by endeavor that could not be obtained by mono-disciplinary activity. Ensuing chemistry- and biology-related project collaborative work called for novel combinatorials and also generated fresh combinatorials in the form of questions, devices, and information. In the course of projects, Gimzewski eloquently expressed in NSR the new disciplinarity criteria of displacement and multiple temporality, while staying in his referent discipline.

Gimzewski began his research career studying properties of nuclear materials. With the development of the STM and soon thereafter the AFM, he immediately adopted these instruments as the bases of his investigations. During the mid and late 1980s, he conducted research on the electronic operations of the STM. In parallel with this, he employed his STM for the study of properties of almost any material that came to hand, but in particular substances of the low-dimensional materials revolution, such as fullerenes. Gimzewski also quickly used the STM to explore the conformation and properties of single molecules. The device gave access to size, shape, position, and to electronic, magnetic, adsorptive, and absorptive forces. Gimzewski stressed the new capacity to manipulate single molecules and atoms, and as a consequence to explore unknown relations and to obtain novel functionalities and functions. He is the co-laureate of the Feynman Prize with Christian Joachim in 1997 (see Chapter 2), for their work using scanning probe microscopes to manipulate molecules.

During the 1980s and 1990s, investigations of the operation of the AFM and its use in physical research gradually emerged as one key theme of Gimzewski's work. It quickly became clear that for him, that domain of the physics discipline associated with mechanical properties lay at the center of his interests. Indeed the AFM comprised a foremost device for exploring mechanical properties on the nanoscale. As indicated in Chapter 1, mechanical activation is the foundational feature of the AFM.[19] Interactions between the object of research and the AFM detection tip are mechanically transmitted to the cantilever, and it is the mechanical motion of the cantilever that registers the intended physical characteristics of the target substance. Mechanical motion thus comprises the logic of the AFM, and the AFM often privileges exploration of mechanical expressions of matter. This physics-grounded complex became the cornerstone of much of Gimzewski's research beginning in the mid-1990s, and it remains so today.

Grounded on his interests in micromechanical instruments relevant for the study of phenomena on the nanoscale, in 1998 Gimzewski developed a thermo-gravimetric instrument five orders more sensitive than any existing apparatus.[20] Pushing the micromechanical approach forward, and expanding the range of mechanical properties that could be accessed at the nanoscale, Gimzewski quickly strove to enhance the sensitivity of mechanical measurement. He saw that the mechanics of vibration versus the simple bending of a cantilever could offer greater nanometric accuracy and could also open his mechanical perspective to the exploration of a broader frame of materials. This switch to vibrational/resonance technology marked the beginning of a chain of transformations that generated ever more combinatorials, where each individual combinatorial often ultimately constituted an element in a more complex combinatorial. It was in part this concatenation that would with time both stimulate Gimzewski to look beyond his discipline for relevant new projects, and function as a resource enabling him to communicate with practitioners from other fields and eventually to initiate projects. In the whole of this work, Gimzewski operated almost exclusively from deep inside his referent physics discipline.

In the latter period of his career, Gimzewski moved intermittently toward the borderland of his referent physics discipline in the course of engaging his projects with

[19] For example, W.Z. Rong, A.E. Pelling, A. Ryan, J. Gimzewski, S.K. Friedlander (2004) Complementary TEM and AFM force spectroscopy to characterize the nanomechanical properties of nanoparticle chain aggregates, *Nano Letters*, 4(11): 2287–2292.

[20] J.H. Fabian, R. Berger, H.P. Lang, C. Gerber, J. Gimzewski, J. Gobrecht, E. Meyer, L. Scandella (2000) Micromechanical thermogravimetry on single zeolite crystals, *Micro Total Analysis Systems '98* Se Mesa Monographs Ct 3rd International Symposium on Micro-Total Analysis Systems (mu-TAS'98) CY OCT 13–16, 1998 CL BANFF, CANADA :117–120.

other fields. Through this, he expressed his new disciplinarity profile. When Gimzewski engaged in projects involving non-physics materials or questions, connections with his referent cognitive and instrumentation home coordinates were always visible and paramount. As his investigations and his instrumentation blossomed, Gimzewski increasingly looked to new domains in which to solve problems, to ask questions, and to take measurements. This did not entail acquiring such exhaustive cognition of an additional discipline as to become a member—or even to wish for or acknowledge a hybrid designation such as physical chemist, biophysicist, etc. By entering into cross-borderland projects, Gimzewski sought fresh territories for expression of his metrological and cognitive orientation.

One of the earliest materializations of the cantilever deflection-sensing device was undertaken in conjunction with work in the discipline of chemistry. An instrument was developed for identification of a range of substances in the form of gases or vapors. Eight cantilevers were coated with alternative metals. The metals contracted at known amounts in reaction with the deposition on their surface of specific substances; and the contraction resulted in the predicted bending of the cantilever. In this fashion, mechanical properties served to observe and specify particular chemical substances and reactions. Between 1997 and 2007, Gimzewski participated in cross-borderland projects associated with a range of chemistry issues over half a dozen times. Such collaborations were only intermittent so they did not lead the physicist to loosen his strong, persistent connections with his home referent. In this sense, the physicist and chemists continued to operate in their respective frames, each speaking his own language. However, cooperation stimulated a fresh combinatorial, common to both parties, in the guise of a new family of device capable of addressing old problems in an entirely new way: chemo-mechanics.

Gimzewski's collaboration with biology-related research, at first sight far from his physics investigations, constitutes a dramatic illustration of projects that temporarily draw scientists far away from the center of their home discipline. In 2000, he engaged in projects with scientists working in the life sciences. One article that reported on the work was entitled "Translating biomolecular recognition into nanomechanics."[21] It makes specific the physicist's mechanics-based orientation to the description and analysis of a species of materials that are alien to physics and that lie at the center of biology-related problems. For this and similar projects, the effectiveness of collaborations required Gimzewski to acquire at least a minimum understanding of the questions asked outside of his discipline, and what they implied in terms of experimental design and operations. For this, he participated in projects that featured medical doctors or specialists in DNA, cell biology, etc.[22] Gimzewski's participation was motivated by the desire to better comprehend vibrational phenomena; the biologists were for their part

[21] J. Fritz, M.K. Baller, H.P. Lang, H. Rothuisen, P. Vettiger, E. Meyer, H.J. Guntherodt, C. Gerber, J. Gimzewski (2000) Translating biomolecular recognition into nanomechanics, *Science*, 288(5464): 316–318.

[22] J. Reed, C. Hsueh, M.L. Lam, R. Kjolby, A. Sundstrom, B. Mishra, J. Gimzewski (2012) Identifying individual DNA species in a complex mixture by precisely measuring the spacing between nicking restriction enzymes with atomic force microscope, *Journal of the Royal Society Interface*, 9(74): 2341–2350.

motivated by the possibility of exploring older questions of membrane elasticity in the novel perspective of mechanical manifestations. Combinatorials came from both sides of the borderland.

Between this initial time and 2012, Gimzewski published over a score of articles resulting from physics–biology interactions. For example, in 2009, he co-authored an article with in vivo embryo development specialists.[23] His perspective permitted the innovative investigation of the properties of stiffness and indentation of a zebra fish (an important model organism) embryonic eye at the nanoscale. In this project, the biologist did not become expert in vibrational phenomena and the physicist did not become an expert in embryology. Nevertheless, the project-based combinatorials allowed each partner to progress in his specific domain thanks to input from neighboring disciplines. Note here that, while Gimzewski published plentifully in this decade in biology-related projects, he circulated multiple times between the heartland of mechanics as physics and outlying borderlands.

Perhaps the most spell-binding bioproject in which Gimzewski has participated and one that he frequently presents, is described by him as "screaming cells."[24] The relevant associated trans-borderland communication and cooperation took place between physicists and cell biologists (cancer, cell membrane, etc.).[25] One goal of the collective work consists of the characterization of mechanics of cell walls in in vivo organisms in a perspective of cancer diagnosis and treatment. Gimzewski's contribution showed modifications in cell wall elasticity which were revealed through their vibrational properties. He could translate these vibrations into acoustical signals, which he refers to as "screaming." Through this project, cell biologists and cancerologists came to reflect on cell behavior and pathology in a novel mechanical perspective. For his part, Gimzewski learned to adapt his instrument to a new environment. Herewith he enriched the repertory of what he included as meaningful signals and in so doing extended his landscape of intelligibility.

We observe that Gimzewski's engagement in the new disciplinarity is linked to his involvement in nanoscale scientific research sparked by the invention of the STM and AFM in the mid-1980s. Prior to these possibilities of combinatorials, his research projects and questions were confined to the pale of traditional disciplinarity.

6.2.3 Shared territory for answering homeland questions

The concepts of combinatorial, project, and referent are strongly present in the following episode. Scientists operating, for example, in the same discipline yet in different specialties

[23] J. Reed, S. Ramakrishnan, J. Schmit, J. Gimzewski (2009) Mechanical interferometry of nanoscale motion and local mechanical properties of living zebrafish embryos, *ACS Nano,* 3(8): 2090–2094.

[24] Interview of Jim Gimzewski (UCLA) by Terry Shinn, 27 January 2008.

[25] S.H. Sharma, C. Santiskulvong, L.A. Bentolila, J.Y. Rao, O. Dorigo, J. Gimzewski (2012) Correlative nanomechanical profiling with super-resolution F-actin imaging reveals novel insights into mechanisms of cisplatin resistance in ovarian cancer cells, *Nanomedicine-Nanotechnology Biology and Medicine,* 8(5): 757–766.

may travel to their respective borderlands to establish a common project. Their aim may be to find a vocabulary and shared experimental objects that will allow them to cast a fresh light on their different homeland-based work. Novelty in the form of a joint insight may emerge from this kind of specialty interaction. Scientists engaged in such projects gain in the framework of their domain.

The case of the two physicists, related in Chapter 2, shows a collaboration between a specialist in optics, Bernard Jusserand, a specialist in acoustics, Bernard Perrin, and an epitaxior, Aristide Lemaître, who supplied the two physicists with semiconductor samples and tailored for them complex nanocavities that they could use for studying both light waves and sound waves. This combination is emblematic of project/discipline displacement dynamics in NSR. The optics practitioner, Jusserand, was experienced in the propagation of optical waves in nanocavities. For his part, Perrin, the acoustical expert, strove to develop semiconductor high-frequency vibrational systems and to understand the dynamics of these vibrations.[26] Working together, they could explore the complex relations between the two expressions of waves. In some of their common experiments, these two categories of waves occur together, and a family of highly original and complex phenomena is generated.

As we have shown in Chapter 2, the two physicists could see, in their collaboration and in the elaboration of a common project, a possibility to develop their own questions and to produce interesting results in this innovative new combined area. The combinatorials between the two specialties were fertile enough to give rise to a summer school, "Son et Lumière."

Jusserand and Perrin's common project occupied an important place in their two careers. Over a period of eight years, they together published 24 articles,[27] but the majority of their respective publications are in their own original field—for example Jusserand, who has published a total of 67 articles since 2004, the date of his encounter with Bernard Perrin, has published 43 without Perrin on his own problematics of research, playing with concepts and hypotheses totally anchored in his specialty referent.[28] This means that, even if one can say that the original idea that joined them was fruitful and that their collaboration has lasted and is still vivid, both of them have continued to work in the heartland of their respective initial home domains of research. A crucial element should

[26] For example: P. Ruello, S. Zhang, P. Laffez, B. Perrin, V. Gusev (2009) Laser-induced coherent acoustical phonons mechanisms in the metal-insulator transition compound $NdNiO_3$: Thermal and nonthermal processes, *Physical Review B*, 79(9); A. Amziane, L. Belliard, F. Decremps, B. Perrin (2011) Ultrafast acoustic resonance spectroscopy of gold nanostructures: Towards a generation of tunable transverse waves, *Physical Review B*, 83(1).

[27] For example: N.D. Lanzillotti-Kimura, A. Fainstein, B. Perrin, B. Jusserand, O. Mauguin, L. Largeau, A. Lemaitre (2010) Bloch oscillations of THz acoustic phonons in coupled nanocavity structures, *Physical Review Letters*, 104(19); A. Huynh, B. Perrin, B. Jusserand, A. Lemaitre (2011) Terahertz coherent acoustic experiments with semiconductor superlattices, *Applied Physics Letters*, 99(19).

[28] For example: C. Aku-Leh, F. Perez, B. Jusserand, D. Richards, G. Karczewski (2011) Dynamical corrections to spin-wave excitations in quantum wells due to Coulomb interactions and magnetic ions, *Physical Review B*, 83(3); W. Peng, F. Jabeen, B. Jusserand, J.C. Harmand, M. Bernard (2012) Conduction band structure in wurtzite GaAs nanowires: A resonant Raman scattering study, *Applied Physics Letters*, 100(7).

be recognized: even though the two specialties, optics and acoustics, are spheres of the scientific discipline physics, they are in fact sufficiently distant from one another to belong to different cognitive worlds evincing distinct referents. To build a common project of research, Jusserand and Perrin had to stand on the respective borderlands of their specialties and to shout over the disciplinary dividing walls. They had to displace from the center of their domain to the periphery; in so doing, they acquired a sense of different temporalities: the temporality of their collaboration, and the temporality of their long-standing work in their home specialty, this feeling probably reinforced by the fact that both of them were experiencing a dynamical "elastic" movement back and forth between the topics, as they worked together and studied separately. These displacements can also be thought of in the framework of respiration, as was defined in Chapter 5. For Jusserand and Perrin, respiration characterizes this very moment when, for distinct reasons and in very different cognitive contexts, they see the opportunity to ask new questions in their respective fields of research, to launch new hypotheses, and to answer questions that would be inaccessible outside of such a collaboration. In this way, one can appreciate their investment in a collaboration as a means of pursuing their own line of investigations. But it also reveals an authentic attempt to understand the other scientist's questions, his logic of thinking. In fact, the effort to displace toward the borderland between their two specialties entailed readjusting their different languages and conceptual referents.

As we saw in Chapter 2, one of the main reasons for the two scientists to join together in a common project was the necessity to work with nano-tailored objects. This implied collaboration with the epitaxior, Aristide Lemaître, who could produce samples and modify them in response to the two scientists' demands, and as their research progressed. Such a collaboration also highlights another important emblematic characteristic of NSR: the central place of control. Here control was crucial in the epitaxy part of the project, where, for example, nanocavities were introduced with Bragg mirrors; but it was also paramount in the experiments themselves, where light and sound impulses were precisely generated and then measured. In this example, where two scientists from different horizons work together on a common project, we see a complex combinatorial of intellectual, instrumental, and material resources. This entailed reframing their respective questions, and organizing multiple discussions in order to generate a fruitful dialog.

6.2.4 Successive projects

As the subtitle suggests, projects are pre-eminent, but this is not to mitigate the continued centrality of referent. In Shimon Weiss's trajectory, which we presented in Chapter 5, the continuity of his work is largely given by the search for ever newer systems of detection in different environments. Here the combinatorials and projects in which he is engaged are cumulative.

The growth of the combinatorials that were involved in his research projects expresses well the necessity for Weiss to federate multiple collaborations from different domains and disciplines to achieve his research aims. In Weiss's case, indeed, his research projects not only progressively entailed the collaboration of ever newer elements (instruments, specialists of other disciplines such as biology and medicine), but also stretched him toward the outer envelope of his specialty, into the borderland between his home referent discipline and other participants. The underlying orientation that characterizes Weiss's trajectory, and constitutes a signature of nano, entails a series of items: the emission/detection of signals emanating from nanoscale objects, the description of single objects, the deterministic perspective in their study, the persisting efforts to control these objects, their links to one another, and their displacements. These various parameters constitute some of the motors that have pushed Weiss to find new combinatorials and to construct new projects in collaboration with practitioners from different specialties; making explicit how decidedly nanoscience research is positioned in the new disciplinarity.

In fact, from Weiss's point of view, the successive projects that he has undertaken can be seen as a chain of questions stemming from the same crucial interrogation and which could only be solved with the necessary collaboration of combinatorials from other disciplines and skills. Each one of the participants, instruments, etc. was considered by him as converging toward the goal that he had decided—the other researchers rooted in other disciplines having their own goals. Stated differently, each practitioner remained firmly attached to his own discipline, yet temporarily participated in, in the instance of Weiss's work, the novel biomolecule project before drifting back to his central disciplinary coordinates.

At each step, additional specialties had to enter the borderland of the research project in order to advance investigations. Today one of Weiss's important endeavors is carried out jointly with a scientist also identified with physics, yet whose recent trajectory encompasses biology, through a series of borderland projects.[29]

CONCLUSION

The goal of this chapter has been to identify the position of NSR research in terms of the several different structural landscapes that enshrine the intellectual organization of contemporary science. The landscapes consist of "interdisciplinarity" and, as set forth here, of "traditional disciplinarity" and "the new disciplinarity." Nanoscience and technology is often declared to be "interdisciplinary." Such a claim imputes a certain form of communication, cognition, and organization in which disciplinary referents necessarily disappear, frontiers between disciplines blur, accumulation of multi-disciplinary works spawns questions having no disciplinary anchor.[30] None of the nanoscientists evoked in

[29] Interview of David Bensimon (Director of the Laboratory of statistic physics in Ecole Normale Supérieure Paris) by authors, 12 May 2009.

[30] R. Pei, A.L. Porter (2011). Profiling leading scientists in nanobiomedical science: Interdisciplinarity and potential leading indicators of research directions, *R&D Management*, *41*(3), 288–306. A.L. Porter, I. Rafols (2009) Is science becoming more interdisciplinary? Measuring and mapping six research fields over time, *Scientometrics*, 81(3): 719–745.

this chapter, and indeed throughout this book, illustrates this profile. None of them has become detached from the disciplinary referents in which they were trained. The referent constitutes their anchor in a particular domain. Borderlands and projects, for their part, express the path by which these scientists elaborate a common project, while not crossing the frontier of their discipline. The notions of displacement and temporality exhibit the short-term period during which these projects are pursued, compared with the long temporality of the research that underpins their career.

Based on sociological literature, and in particular scientometric studies, the connection between NSR and interdisciplinarity is unclear. Some argue that the field is interdisciplinary to the same extent as all other domains are evolving in that direction. Others insist that NSR exhibits no traces of interdisciplinarity.[31] A few consider that it is cognitively interdisciplinary, but not organizationally so.[32] One thing traverses these opposing claims: little attention has been paid to clearly defining what is intended as "interdisciplinary." Perhaps the best approximation revolves around the concept of "cognitive integration."[33] What this stimulating term exactly covers is yet to be determined, but one thing is certain, no kind of fusional, totally interpenetrating, undifferentiated knowledge can be found in the cognitive productions of the NSR scientists that are reported in this chapter. If interdisciplinarity successfully characterizes the endeavors of nanopractitioners, the ways in which this is so are yet to be elucidated.

Nevertheless one thing is sure: expressions of traditional disciplinarity persist in many areas of science, for example organic and inorganic chemistry, acoustics, seismology, and geology. Here the disciplinary referent is closed and inward looking, and few mechanisms exist that promote trans-borderland communication. Our interviews of many NSR scientists have yielded only a handful of investigators who correspond with this profile. In contrast, NSR scientists largely operate along lines compatible with the structures and dynamics of the new disciplinarity.

Three of the four case studies presented in this chapter emphasize particular expressions of the new disciplinarity trajectories, where a constant remains the disciplinary referent. The case of Fraser Stoddart ("Beckoning from in-doors") is emblematic of traditional disciplinarity in NSR. This attachment to traditional forms of discipline is, as already suggested, extremely rare. All through his scientific career, he has ploughed the same soil of questions inside his disciplinary framework. He stands fully in the pattern of

[31] J. Schummer (2004) Multidisciplinarity, interdisciplinarity, and patterns of research collaboration in nanoscience and nanotechnology, *Scientometrics*, 59(3): 425–465.

[32] I. Rafols, M. Meyer (2007) How cross-disciplinary is bionanotechnology? Explorations in the specialty of molecular motors, *Scientometrics*, 70 (3): 633–650; I. Rafols, M. Meyer (2010) Diversity and network coherence as indicators of interdisciplinarity: Case studies in bionanoscience, *Scientometrics*, 82(2): 263–287; A.L. Porter, I. Rafols (2009) Is science becoming more interdisciplinary? Measuring and mapping six research fields over time, *Scientometrics*, 81(3): 719–745.

[33] A.L. Porter (2009) How interdisciplinary is nanotechnology? *Journal of Nanoparticle Research*, 11(5): 1023–1041; A.L. Porter (to appear *R&D*) Management profiling leading scientists in nano-biomedical science: Interdisciplinarity and potential leading indicators of research directions; Ruimin Pei Chinese Academy of Sciences School of Public Policy Georgia Institute of Technology.

traditional disciplinarity, acquiring there an ever more complete control of geometrical configurations and intertwinings of complex molecules. Beckoning from in-doors, Stoddart made his molecular models easily understandable to specialists in other disciplines. His ambition was thus to make practitioners in other fields become interested in his findings and personal perspective.

For its part, the trajectory of James Gimzewski ("Extending the repertory") represents the new disciplinarity, particularly as regards "displacement." Through different instrumental combinatorials and through temporary collaborations with scientists from other domains in physics, chemistry, and even biology, he progressively refined his interests and competence on nanoscale mechanical vibrational phenomena. Little by little he changed his object of investigation, becoming increasingly interested in biological materials. One can say that these temporary displacements, progressively both more deeply inscribed him in the investigation of vibrational mechanical phenomena, and made his competence in this domain more and more susceptible to building projects, in different combinatorials with other specialties.

In our third case study, the trajectory of two physicists working in two different specialties (optics and acoustics) developed "Shared territory for answering homeland questions." Each scientist remained inside his homeland with the disciplinary referents intact. But in order to build a common project, they had to evolve toward their respective disciplinary borderlands. Each of them sustained his questions and aims. The borderlands represent the zones where they sought to understand the other's language, logic, and questions. The common project made it possible for each scientist to go further in his respective research problematics and to benefit from more complex cognitive and instrumental combinatorials. In that sense, the new disciplinarity does not erode frontiers between domains, nor does it mitigate scientific questions belonging to a specialty. It makes visible how collaborations function between domains. In this way, the new disciplinarity elucidates the importance of the role played by the borderland of each specialty.

Shimon Weiss's trajectory ("Successive projects") illustrates another important characteristic of the dynamic that new disciplinarity entails. Through his successive multiple nanoprojects, and progressively adapting his methods, questions, instruments, with novel combinatorials, Weiss evolved in his work. Little by little, his interest passed from pure physics questions of emitting/detecting fluorescence phenomena to the deterministic perspective of controlling single molecules in increasingly complex milieus (in vivo biological environments). He thus changed over time as he evolved toward projects which are ever more distant from his primary questions in physics. But a closer inspection of his trajectory reveals that, even though the different projects he undertook during this evolution incorporated scientific specialties and skills of very diversified horizons, each scientist in the common projects nevertheless remained firmly in his own area of questions and interests, and in so doing, each one continued to bring to the others the spirit and the richness of his own home discipline.

In conclusion, although our investigations suggest that it is appropriate to describe NSR in the framework of what we have termed "the new disciplinarity," much additional work is required to see clearly how interdisciplinarity, traditional disciplinarity, and new disciplinarity play out in the future. It is early days for NSR. The domain came into being only about three decades ago. When compared with many other fields, it is still an infant. The structures and dynamics of NSR are perhaps not yet fully set. Moreover, across history, the substance and evolution of the parameters of the organization of cognition shift, and this will certainly apply to interdisciplinarity, the new disciplinarity, and traditional disciplinarity. Time will tell.

General Conclusion

The interviews and observations undertaken in the course of our research project have revealed that science at the nanoscale carries with it a particular way of observing, describing, and interpreting. It similarly vehicles new scientific questions and ways of formulating them, a novel approach to research materials, and finally unprecedented orientations in the organization of scientific activity. The advent of what can be viewed as an instrument revolution, be it metrological or numerical and their combinatorials, paralleled by the synthesis of new families of substances, sparked a nanoscience research "gold rush" between 1985 and 1995. Hundreds of scientists swarmed to NSR and the number of publications in related fields rocketed.

The study of the physical properties of single molecules lies at the heart of NSR. Astonishingly, scientists can "see" surface features of molecules in the form of mountains, plains, and valleys in much the same way that gazing through his telescope Galileo explored and described the face of the moon four hundred years ago. At the nanometer scale, molecular objects exhibit different and indeed entirely novel categories of properties not to be found in the bulk materials of nature. What exactly has made observation of surface features of individual molecules possible? Exploration of this world has been contingent on the mobilization and integration of two main families of instruments—scanning probe microscopy and numerical simulation. These instruments combine to form an NSR exploratory system and moreover, they successfully interface with other species of devices. This is an expression of what we term "combinatorials," which is an underlying characteristic of NSR. Combinatorials stretch far beyond instrumentation: they equally include methods, concepts, materials, questions, skills, etc. Because of the frequency and concentration of combinatorials in NSR, they may be regarded as constituting one of the bases of the domain. Of foremost importance, NSR's foundational instruments introduce an additional key element of the field. They enable and generalize the capacity for manipulation at the molecular level. For example, the scanning tunneling microscope tip can not only read the surface landscape of molecules and atomic clusters, but also can displace them. Individual atoms and molecules can be positioned at will. In this case it functions as a mechanism of control.

Through control, universes of new substances have been invented. Synthesis is one of the two mainstays of NSR along with instrumentation. Materials such as fullerenes, self-assembled monolayers (SAMs), pre-designed synthesized biomolecules (like custom made proteins or "DNA origami"), and semiconductor low-dimensional objects are examples of NSR artificial substances. Synthesis of these families of materials has given rise to the birth and development of a specialty inside nano, namely epitaxy. In NSR, metrology experimenters call for specifically tailored research-objects (objects by design) and the task of epitaxiors is to materialize the demand. This relationship between experimentation and epitaxy introduces an important epistemological issue. In most instances, historically, researchers formulated questions with reference to the objects extant in nature. In NSR, however, the direction of the epistemological arrow between materials and question is often reversed. The capacity to fashion materials frees practitioners to elaborate questions, unfettered by the constraints of nature. In effect, substances are now designed in response to scientists' queries about physical properties.

Indeed, the new potential of almost infinite materials permutations, the connected capacity for the design of objects, and their embodiment through synthesis and subsequent study, are particularly important in nanobiology. Biological materials entered nanoresearch mainly through the initiatives of physicists who in the early days cast about in all directions for new families of substances to explore with their new apparatus. In NSR, biologists saw opportunities for novel information and insight into their matter and they found new tools to manipulate and control this—proteins and DNA. In the perspective of nano, biologists have obtained, as never before, measurements of, and insights into the relations between structure / form and function, binding, etc. The capacity for control that is characteristic of nano has given rise to the ability to design and build new synthetic nanobiological objects for purposes of experimentation and broader application.

Nanobiology investigation has additionally opened a window into the exploration of processes and dynamics. The important dynamical events present in biological molecules can now be precisely monitored and even "visualized." In view of the incomparable intricacy and complexity of biomolecules, the horizons for investigation and the existence of uncharted possibilities is almost incalculable. It is this that has stimulated the appetite of many physicists and chemists for research employing nanobiological materials. The tailoring of objects at the nanoscale, be it biomaterials, SAMs, or semiconductors, in response to scientists' problematics, often entailing the creation of specific morphologies, largely relies on imagery.

Five epistemological orientations frame NSR: (1) image, (2) form, (3) local, (4) descriptivism, (5) determinism. Today, wherever one looks inside scientific research, information in the form of images tends to be omnipresent. Imagery may in part be seen as a general cultural evolution that indeed extends well beyond science. Since the earliest days of NSR, visual representations emerged as a preferred expression of data. Metrological and numerical apparatus both depict physical and biological objects as three-dimensional images. They are interpreted by scientists as somehow conveying the physical materiality of substances. They are not merely a vehicle of communication, but also a basis for complementarity or confrontation between simulation and metrological results, analysis, discussion, and interpretation. Indeed, images are a first epistemological component of NSR.

The importance of imagery in NSR reflects the centrality of form as the understanding of physical phenomena at the nanoscale. Scientists describe and think in terms of spatial coordinates and these are best apprehended as forms, and as forms exhibited through imagery. In nano, form promotes understanding of not only architecture but also many physical properties. Nano has indeed developed a specialized vocabulary of form which includes form as morphology and as spatial occupancy, force, and perturbation. Form is constitutive of practitioners' reasoning, and their formulation of questions which makes it an integral part of nano epistemology. Investigation of form at the nanoscale necessarily situates reflection on the local. Scientists concentrate on the singular and intra-singular or short distance interrelations between entities. The referent is thus the nanoscale event proper and its immediate environment. Broader landscapes are excluded. In this perspective, modeling and theory are seldom relevant. The local, enjoys an epistemological status in NSR.

Indeed, the conjunction of form and local is constitutive of research results, and when taken together they are the engine of description. It is fair to say that NSR is a descriptivist science. It consists of successively pointing out aspects of objects and the connections between them. It may be compared with a kind of narrative. In some ways, this harks back to the late nineteenth century when science was rich with description—an age of descriptivism. Descriptivism constitutes a fourth epistemological feature of NSR.

Historically, molecular phenomena have long been studied in a stochastic perspective. This is the signature of quantum physics and chemistry. In contrast, NSR is deterministic: the investigative tools of the field now make it possible to precisely observe and appreciate position and shape of individual nanometer objects. Moreover, scientists can select and place single, stable molecules at will, and then move them about and connect them according to plan—in a word, control. What could possibly be more characteristic of determinism? Nano is certainly not the realm of probability.

On a related epistemological level, our study indicates that NSR has contributed little to the domain of theory. Theory as abstraction, integration, and generalization is by and large absent from NSR. It appears only in an oblique way. It naturally occurs as background knowledge, but it does not lie at the forefront of metrological and simulation work. On a different register, simulation frequently generates predictions that are very important in NSR. One function of prediction is linkage between simulation and metrological results. It is the predictions and explanations offered by simulation that nourish reflection on the possible connections that might exist between the metrological findings occurring in a specific class of materials, or with reference to a category of properties. In a word, NSR's emphasis on the singular, the local, form, and description which preoccupies questions, thinking, and findings of practitioners largely accounts for the relegation of things theoretical to the background. Whether this will change with the maturing of NSR is something to be observed over time.

This book points to two aspects of NSR practitioners' socio-cognitive trajectories. These aspects are expressed as "respiration" and the "new disciplinarity." Respiration describes patterns of continuity and discontinuity on practitioners' selection of and engagement with successive intellectual and technical communities and problematics. We have demonstrated that in NSR, scientists characteristically intermittently take stock of the huge number of instruments, materials, and thematic possibilities that underpin nano investigation, and in view of the sweep of cognitive and instrumental resources available to them, they choose to shift directions or to carry on their current work. Respiration is a cognitive driven standard in NSR which goes well beyond the historically observable processes of professional mobility. Indeed, respiration counts as one of the internal transformative mechanisms that fuels nano.

Respiration in research activity consists of those in-between project intervals in a scientific trajectory when a scientist evaluates his present and likely future intellectual possibilities, and considers ensuing research options and the best path in view of mainly cognitive materials and technical criteria. In NSR, the practice of respiration has given rise to several geometries of research structure and strategy. Respiration functions inside what we have called the "concentration sphere," where epistemological expectations are paramount. Here, respiration represents moments during research processes when practitioners reflect on the potential of their combinatorial resources, in terms of continuing a project or shifting to a new set of questions. Beyond the concentration sphere, some NSR practitioners engage trajectories that take them into what we call the "extension sphere." The scope of combinatorials inside this sphere is both different and broader than in concentration. The research framework of extension is the "function-horizon," where research is structured in terms of attaining practical outputs as opposed to epistemological ends. In this environment, combinatorials often include economic, organizational, bureaucratic, and ideological components alongside more technical ones.

NSR is indeed a young science: along what lines is it organized and does it operate? In the lively contemporary debate on interdisciplinarity versus disciplinarity, our investigations suggest that NSR scientists frequently operate in a framework of a variant kind of disciplinarity, which we term the "new disciplinarity." In the new disciplinarity, disciplines continue to comprise the principal referent of practitioners' concepts and research questions—the signatures of a scientist's home discipline. On the other hand, such disciplines promote much intra-discipline mobility, where practitioners circulate back and forth between the center and periphery. Because of the vast quantity of cognitive, instrumentation, and materials combinatorials available in NSR (a veritable culture of combinatorials), scientists formulate questions that cannot find answers strictly within the confines of one discipline. Individuals speak across the borderland of their home discipline to other practitioners standing in the borderland of another discipline, as they pursue a shared project. This organization and dynamic of research work contrasts with the closed, inward looking norms and practices of nineteenth-century traditional disciplinarity. It similarly repudiates the frequently voiced tenet of the demise of frontiers and total cognitive integration espoused by interdisciplinarity. The new disciplinarity, driven by combinatorials, conjoins the discipline's structure of referent with internal elasticity and intermittent circulation toward borderlands, thus promoting research with colleagues from other disciplines who likewise sustain their disciplinary referent.

Finally, this investigation of central aspects of nanoscale scientific research invites the question: in what ways and to what extent does NSR constitute a kind of "scientific revolution"? It is still early days and the verdict is not yet in. Whatever the reply, one thing is clear: NSR represents an undisputed transformation of key components of contemporary science.

REFERENCES

Abir-Am, P.G. (2006) Molecular biology and its recent historiography. A transnational quest for the "Big Picture," *History of Science*, 44: 95–118.

Abbott, A. (1995) Things of boundaries, *Social Research*, 62(4): 857–882.

Abbott, A. (2001) *Chaos of Disciplines*. Chicago and London: The University of Chicago Press.

Aigouy, L., Y. de Wilde, C. Frétigny (2007) *Les Nouvelles Microscopies. A la découverte du nanomonde*. Paris: Belin.

Akahane, Y., R. Asano, B.S. Song, S. Noda (2003) High-Q photonic nanocavity in a two-dimensional photonic crystal, *Nature*, 425(6961): 944–947.

Aku-Leh, C., F. Perez, B. Jusserand, D. Richards, G. Karczewski (2011) Dynamical corrections to spin–wave excitations in quantum wells due to Coulomb interactions and magnetic ions, *Physical Review B*, 83(3).

Alberts, B., A. Johnson, J. Lewis, M. Raff, K. Roberts, P. Walter (2002) *Molecular Biology of the Cell*. New York, NY: Garland (4th ed.).

Alivisatos, P. (2004) The use of nanocrystals in biological detection, *Nature Biotechnology*, 22(1): 47–52.

Allamel-Raffin, C. (2004) *La Production et les fonctions des images en physique des matériaux et en astrophysique*. Doctoral dissertation, University of Strasbourg.

Allamel-Raffin, C. (2006) La Complexité des images scientifiques. Ce que la sémiotique de l'image nous apprend sur l'objectivité scientifique, *Communication et langages*, 149(1): 97–111.

Altschul, S.F., W. Gish, W. Miller, E.W. Myers, D.J. Lipman (1990) Basic local alignment search tool, *Journal of Molecular Biology*, 215(3): 403–410.

Altschul, S.F., T.L. Madden, A.A. Schaffer, Z. Zhang, W. Miller, D.J. Lipman (1997) Gapped BLAST and PSI-BLAST: A new generation of protein database search programs, *Nucleic Acids Research*, 25(17): 3389–3402.

Amabilino, D.B., J.F. Stoddart (1995) Interlocked and intertwined structures and superstructures, *Chemical Reviews*, 95(8): 2725–2828.

Amziane, A., L. Belliard, F. Decremps, B. Perrin (2011) Ultrafast acoustic resonance spectroscopy of gold nanostructures: Towards a generation of tunable transverse waves, *Physical Review B*, 83(1).

Antoshchenkova, E., M. Hayoun, F. Finocchi, G. Geneste (2012) Kinetic Monte-Carlo simulation of the homoepitaxial growth of MgO{001} thin films by molecular deposition, *Surface Science*, 606(5–6): 605–614.

Apker, L., E. Taft (1950) Evidence for exciton-induced photoelectric emission from f-centers in alkali halides, *Science*, 112(2911).

Aviram, A., C. Joachim, M. Pomerantz (1988) Evidence of switching and rectification by a single molecule effected with a scanning tunneling microscope, *Chemical Physics Letters*, 146(6): 490–495.

Avouris, P. (1990) Atom-resolved surface-chemistry using the scanning tunneling microscope, *Journal of Physical Chemistry*, 95(6): 2246–2256.

Avouris, P., D.E. Demuth (1981) Spectroscopy and photophysical dynamics of aromatics adsorbed on a clean Ag(111) surface, *Journal of Photochemistry*, 17(1–2): 111–112.

Avouris, P., T. Hertel, R. Martel (1998) Manipulation of individual carbon nanotubes and their interaction with surfaces, *Journal of Physical Chemistry B*, 102(6): 910–915.

Avouris, P., R. Wolkow (1989) Scanning tunneling microscopy of insulators – Caf2 epitaxy on Si(111), *Applied Physics Letters*, 55(4): 1074–1076.

Baba, Y., G. Dujardin, P. Feulner, D. Menzel (1991) Formation and dynamics of exciton pairs in solid argon probed by electron-stimulated ion desorption, *Physical Review Letters*, 66(25): 3269–3272.

Baird, D. (2004) *Thing Knowledge. A Philosophy of Scientific Instruments*. Berkeley: University of California Press.

Baird, D., A. Shew (2004) Probing the history of scanning tunneling microscopy. In: D. Baird, A. Nordmann, J. Schummer (eds.) *Discovering the Nanoscale*. Amsterdam: IOS Press, pp.14–156.

Baird, D., A. Nordmann, J. Schummer (eds.) (2004) *Discovering the Nanoscale*. Amsterdam: IOS Press.

Ball, P. (2002) *Stories of the Invisible, a Guided Tour of Molecules*. Oxford University Press.

Bas van Fraassen, C. (ed.) (2008) *Scientific Representation: Paradoxes and Perspective*. Oxford: Clarendon Press.

Bellessa, J., R. Grousson, V. Voliotis, X.L. Wang, M. Ogura, H. Matsuhata (1997) High spatial resolution spectroscopy of a single V-shaped quantum wire, *Applied Physics Letters*, 71(17): 2481–2483.

Bellessa, J., V. Voliotis, X.L. Wang, M. Ogura, H. Matsuhata, et al. (1997) Evidence for exciton localization in V-shaped quantum wires, *Physica Status Solidi A-Applied Research*, 164(1): 273–276.

Ben-David, J. (1971) *The Scientist's Role in Society, A comparative study*. Englewood Cliffs, NJ: Prentice-Hall.

Bensaude-Vincent, B. (2001) The construction of a discipline: Materials science in the United States, *Historical Studies in the Physical and Biological Sciences*, 31(2): 223–248.

Bensaude-Vincent, B. (2009) *Les Vertiges de la technoscience: Façonner le monde atome par atome*. Paris: La Découverte.

Berndt, R., J. Gimzewski (1993) Photon-emission in scanning-tunneling-microscopy – Interpretation of photon maps of metallic systems, *Physical Review B*, 48(7): 4745–4754.

Berton, P. (2001) *Klondike: The Last Great Gold Rush (1896–1899)*. Toronto: Kindle Edition.

Binnig, G.H. Rohrer (1993) *Scanning Tunneling Microscopy – From Birth to Adolescence*, Nobel Prize Lecture 8 December 1986. In: *Physics 1981–1990*, Editor-in-Charge Tore Frängsmyr, Editor Gösta Ekspang. Singapore: World Scientific Publishing Co.

Black, M. (1962) *Models and Archetypes. Models and Metaphors, Studies in Language and Philosophy*. Ithaca, NY: Cornell University Press.

Borensztein, Y., L. Delannoy, A. Djedidi, R.G. Barrera, C. Louis (2010) Monitoring of the plasmon resonance of gold nanoparticles in Au / TiO_2 catalyst under oxidative and reducing atmospheres, *Journal of Physical Chemistry C*, 114(19): 9008–9021.

Bottin, F., F. Finocchi, C. Noguera (2005) Facetting and (nx1) reconstructions of $SrTiO_3$(110) surfaces, *Surface Science*, 574(1): 65–76.

Bourdieu, P. (1975) La Spécificité du champ scientifique et les conditions sociales du progrès de la raison, *Science et structure sociale*, 7(1): 91–118.

Bourdieu, P. (1984) *Homo Academicus*. Paris: Minuit.

Bourdieu, P. (2001) *Science de la science et reflexivite*. Paris: Raisons d'Agir.

Bromberg, J.L. (1991) *The Laser in America. 1950–1970*. Cambridge, MA: MIT Press.

Bruchez Jr, M., M. Moronne, P. Gin, S. Weiss, A.P. Alivisatos (1998) Semiconductor nanocrystals as fluorescent biological labels, *Science*, 281(5385): 2013–2016.

Brus, L. (1984) Electron electron–hole interactions in small semiconductor crystallites. The size dependence of the lowest excited electronic state, *Journal of Chemical Physics*, 80(9): 4403–4409.

Bryngelson, J.D., J.N. Onuchic, N.D. Socci, P.G. Wolynes (1995) Funnels, pathways, and the energy landscape of protein-folding – A synthesis, *Proteins-Structure Function and Genetics*, 21(3): 167–195.

Bryngelson, J.D., J.N. Onuchic, N.D. Socci, P.G.Wolynes (1995) Funnels, pathways, and the energy landscape of protein-folding – A synthesis, *Proteins-Structure Function and Genetics*, 21(3): 167–195.

Bueno, O. (2006) Representation at the nanoscale, *Philosophy of Science*, 76: 617–628.

Bueno, O. (2008a) Scientific representation. Microscopes and mathematical models. In: C. Bas van Fraassen (ed.) *Scientific Representation: Paradoxes and Perspective*. Oxford: Clarendon Press.

Bueno, O. (2008b) Visual evidence at the nanoscale, *Spontaneous Generations: A Journal for the History and Philosophy of Science*, 2(1).

Bueno, O. (2011) When physics and biology meet: The nanoscale case, *Studies in History and Philosophy of Science*, Part C, 42(2): 180–189.

Buffon, G. de (1749) *Histoire naturelle, générale et particulière, avec la description du Cabinet du Roy*. 3 Volumes Tome I Texte établi, introduit et annoté par S. Schmitt avec la collaboration de C. Crémière. Paris: Honoré Champion (2007–2009).

Burian, R.M. (1993) Technique, task definition, and the transition from genetics to molecular genetics: Aspects of the work on protein synthesis in the laboratories of J. Monod and P. Zamecnik, *Journal of the History of Biology*, 26: 387–407.

Bush, V. (1945) *The Endless Frontier. A Report to the President by Director of the Office of Scientific Research and Development*. Washington, DC: United States Government Printing Office.

Cambrosio, A., D. Jacobi, P. Keating (2005) Arguing with images, Pauling's theory of antibody formation, *Representations* (Journal of Digital Publishing), 89(1): 94–130.

Cambrosio, A., D. Jacobi, P. Keating (2006) Arguing with images. Pauling's theory of antibody formation. In: L. Pauwels (ed.) *Visual Cultures of Science: Rethinking Representational Practices in Knowledge Building and Science Communication*. Dartmouth, NH: Dartmouth College Press, pp. 153–194.

Campilho, A., M.S. Kamel (eds.) (2006) *Image Analysis and Recognition*. Berlin, Heidelberg: Springer.

Cartwright, N. (1983) *How the Laws of Physics Lie*. Oxford: Clarendon Press.

Cartwright, N. (1991) Can wholism reconcile the inaccuracy of theory with the accuracy of prediction?, *Synthese*, 89(1): 3–13.

Cartwright, N. (1997) Where do laws of nature come from?, *Dialectica*, 51(1): 65–78.

Carusi A., B. Rodriguez, K. Burrage (2013) Models systems in computational systems biology. In: J.M. Duràn, E. Arnold (eds.) *Computer Simulations and the Changing Face of Scientific Experimentation*. Cambridge Scholars Publishing, pp. 118–145.

Chadarevian, (de) S., N. Hopwood (eds.) (2004) *Models. The Third Dimension of Science*. Palo Alto, CA: Stanford University Press.

Chen J.H., N.R. Kallenbach, N.C. Seeman (1989) A specific quadrilateral synthesized from DNA branched junctions, *Journal of the American Chemical Society*, 111(16): 6402–6407.

Chichak, K.S., J. Cantrill, A.R. Pease, S.H. Chiu, G.W.V. Cave, J.L. Atwood, J.F. Stoddart (2004) Molecular borromean rings, *Science*, 304(5675): 1308–1312.

Cho, A.Y. (1999) How molecular beam epitaxy (MBE) began and its projection into the future, *Journal of Crystal Growth*, 201: 1–7.

Choi, H. (2007) The boundaries of industrial research: Making transistors at RCA, 1948–1960, *Technology and Culture*, 48(4): 758–782.

Choi, H., C. Mody (2013) From materials science to nanotechnology: Institutions, communities, and disciplines at Cornell University, 1960–2000, *Historical Studies in the Natural Sciences*, 43(2): 121–161.

Collier, C.P., E.W. Wong, M. Belohradsky, F.M. Raymo, J.F. Stoddart, P.J. Kuekes, R.S. Williams, J.R. Heath (1999) Electronically configurable molecular-based logic gates, *Science*, 285(5426): 391–394.

Cren, T., D. Fokin, F. Debontridder, V. Dubost, D. (2009) Roditchev Ultimate vortex confinement by scanning tunneling spectroscopy, *Physical Review Letters*, 102(12): 74–78.

D'Arcy Thompson, W. (1907) *On Growth and Form*. Cambridge: Cambridge University Press.

Dattelbaum, J.D., L.L. Looger, D.E. Benson, K.M. Sali, R.B. Thompson, H. Hellinga (2005) Analysis of allosteric signal transduction mechanisms in an engineered fluorescent maltose biosensor, *Protein Science*, 14(2): 284–291.

Debru, Cl. (1983) *L'Esprit des protéines*. Paris: Hermann.

Dekker, E. (2012) *Illustrating the Phaenomena: Celestial Cartography in Antiquity and the Middle Ages*. Oxford University Press.

Denis, C.T. Greenbaum (1991) *Image and Cognition*. New York, NY: Harvester Wheatsheaf.

Deniz, A.A., T.A. Laurence, G.S. Beligere, M. Dahan, A.B. Martin, D.S. Chemla, P.E. Dawson, P.G. Schultz, S. Weiss (2000) Single-molecule protein folding: Diffusion fluorescence resonance energy transfer studies of the denaturation of chymotrypsin inhibitor, 2, *Proceedings of the National Academy of Sciences USA*, 97(10): 5179–5184.

Drexler, K.E. (1986). *Engines of Creation*, Anchor. Garden City.

Drexler, K.E. (1992) *Nanosystems: Molecular Machinery, Manufacturing, and Computation*. John Wiley & Sons, Inc.

Dujardin, G., S. Leach, O. Dutuit, P.M. Guyon, M. Richardviard (1984) Double photoionization of SO_2 and fragmentation spectroscopy of SO_2^{++} studied by a photoion photoion coincidence method, *Chemical Physics*, 88(3): 339–353.

Dujardin, G., R.E. Walkup, P. Avouris (1992) Dissociation of individual molecules with electrons from the tip of a scanning tunneling microscope, *Science*, 255(5049): 1232–1235.

Duràn, J.M., E. Arnold (2013) (eds.) *Computer Simulations and the Changing Face of Scientific Experimentation*. New Castle: Cambridge Scholars Publishing.

Ebright, R.H., S. Weiss, A. Chakraborty, D. Wang, Y. Korlann, A. Kapanidis, E. Margeat (2009) Single-molecule analysis of transcription, *Faseb Journal*, 23 Meeting Abstract: 202.1.

Edgerton, S.Y. (1988) Aesthetic and digital image processing representational craft in contemporary astronomy. In: G. Fyfe, J. Law (eds.) *Picturing Power; Visual Depictions and Social Relations*. London: Routledge, pp. 184–220.

Eigler, D.M., E.K. Schweizer (1990) Positioning single atoms with a scanning tunneling microscope, *Nature*, 344(6266): 524–526.

Ertl, G. (2008) Reactions at surfaces: From atoms to complexity (Nobel lecture), *Angewandte Chemie International Edition*, 47(19): 3524–3535.

Espeland, W., M. Stevens (1998) Commensuration as a social process, *Annual Review of Sociology*, 24: 313–343.

Eto, H. (2003) Interdisciplinary information input and output of nano-technology project, *Scientometrics*, 58(1): 5–33.

Fabian, J.H., R. Berger, H.P. Lang, C. Gerber, J. Gimzewski, J. Gobrecht, E. Meyer, L. Scandella (2000) Micromechanical thermogravimetry on single zeolite crystals, *Micro Total Analysis Systems '98* Se Mesa Monographs Ct 3rd International Symposium on Micro-Total Analysis Systems (mu-TAS'98) CY OCT 13-16, 1998 CL BANFF, CANADA: 117–120.

Fabian, D., J. Gimzewski, A. Barrie, B. Dev (1977) Excitation of Fe1s core-level photoelectrons with synchrotron radiation, *Journal of Physics F-Metal Physics*, 7(12): 345–348.

Farge, M. (1990) L'Imagerie scientifique. Choix des palettes de couleur pour la visualisation des champs scalaires bidimensionnels, *L'Aéronautique et l'astronautique*, 1(140): 24–33.

Feltrin, A., K.R. Idrissi, A. Crottini, M.A. Dupertuis, J.L. Staehli, B. Deveaud, V. Savona, X.L. Wang, M. Ogura (2004) Exciton relaxation and level repulsion in GaAs/AlxGa1-xAs quantum wires, *Physical Review B*, 69(20).

Fernandez, A., E.I. Shakhnovich (1990) Activation-energy landscape for metastable RNA folding, *Physical Review A*, 42(6): 3657–3659.

Finocchi, F., R. Hacquart, C. Naud, J. Jupille (2008) Hydroxyl-defect complexes on hydrated MgO smokes, *Journal of Physical Chemistry C*, 112(34): 13226–13231.

Fischer, B., S. Sternklar, S. Weiss (1986) Theoretical and experimental study of frequency detuning in photorefractive oscillators, *Journal of the Optical Society of America A-Optics Image Science and Vision*, 3(13).

Fox, R. (2013) *The Oxford Handbook of the History of Physics*. Oxford: Oxford University Press.

Fraassen (van) Bas C. (2008) *Scientific Representation. Paradoxes and Perspectives*. Oxford University Press.

Francoeur, E. (1997) The forgotten tool: The design and use of molecular models, *Social Studies of Science*, 27(1): 7–40.

Francoeur, E. (2004) From model kits to interactive computer graphics. In: S. de Chadarevian, N. Hopwood (eds.) *Models. The Third Dimension of Science*. Palo Alto, CA: Stanford University Press, pp. 402–433.

Francoeur, E., J. Segal (2004) From model kits to interactive computer graphics. In: S. de Chadarevian, N. Hopwood (eds.) *The Third Dimension of Science*. Stanford, CA: Stanford University Press, pp 402–429.

Frauenfelder, H., S. Sligar, P.G. Wolynes (1991) The energy landscapes and motions of proteins, *Science*, 254: 1598–1603.

Fritz, J., M.K. Baller, H.P. Lang, H. Rothuisen, P. Vettiger, E. Meyer, H.J. Guntherodt, C. Gerber, J. Gimzewski (2000) Translating biomolecular recognition into nanomechanics, *Science*, 288(5464): 316–318.

Frodeman, R., J. Thompson-Klein, C. Mitchan (2010) *The Oxford Handbook of Interdisciplinarity*. Oxford: Oxford University Press.

Fujimura, J.H. (1992) Crafting science: Standardized packages, boundary objects and "translation." In: A. Pickering (ed.) *Science as Practice and Culture*. Chicago, IL: University of Chicago Press.

Fujimura, J.H., M. Fortun (1996) Constructing knowledge across social worlds: The case of DNA sequence databases in molecular biology. In: L. Nader (ed.) *Naked Science. Anthropological Inquiry into Boundary, Power and Knowledge*. London: Routledge.

Fyfe, G., J. Law (eds.) (1988) *Picturing Power; Visual Depictions and Social Relations*. London: Routledge.

Galison, P. (1997) *Image of Logic. A Material Culture of Microphysics*. Chicago, IL: University of Chicago Press.

Galison, P., L. Daston (2007) *Objectivity*. New York: Zone Books.

Gamow, G. (2002) *The Cambridge Dictionary of Scientists*. Cambridge: Cambridge University Press (2nd edition).

Gassman, N., S. Ho, Y. Korlann, J. Chiang, Y. Wu, L.J. Perry, Y. Kim, S. Weiss (2009) In vivo assembly and single-molecule characterization of the transcription machinery from Shewanella oneidensis MR-1, *Protein Expression and Purification*, 65(1): 66–76.

Gavroglou, K., Y. Goudaroulis (1989) *Methodological Aspects of the Development of Low Temperature Physics, 1881–1956: Concepts out of Context(s)*. Dordrecht: Kluwer Academic Publishers.

Gavroglou, K., A. Simões (2012) *Neither Physics Nor Chemistry*. Cambridge, MA: MIT Press.

Geysermans, P., F. Finocchi, J. Goniakowski, P. Hacquart, J. Jupille (2009) Combination of (100), (110) and (111) facets in MgO crystals shapes from dry to wet environment, *Physical Chemistry – Chemical Physics*, 11(13): 2228–2233.

Ghadiri, M.R., J.R. Granja, R.A. Milligan, D.E. Mcree, N. Khaznovich (1993) Self-assembling organic nanotubes based on a cyclic peptide architecture, *Nature*, 366(6453): 324–327.

Gibbons, M., C. Limoges, H. Nowotny, et al. (1994) *The New Production of Knowledge: The Dynamics of Science and Research in Contemporary Societies*. London: Sage.

Giere, R. (2010) *Explaining Science: A Cognitive Approach*. Chicago, IL: University of Chicago Press.

Gimzewski, J., A. Humbert, D. Pohl, S. Veprek (1986) Scanning tunneling microscopy of nanocrystalline silicon surfaces, *Surface Sciences*, 168(1–3): 795–800.

Gimzewski, J., B. Reihl, J.H. Combes, R. Schlittler (1988) Photon-emission with the scanning tunneling microscope, *Zeitschrift fur Physik B-Condensed Matter*, 72(4): 497–501.

Gimzewski, J., E. Stoll, R.P. Schlittler (1987) Scanning tunneling microscopy of individual molecules of copper phthalocyanine adsorbed on polycrystalline silver surfaces, *Surface Sciences*, 181(1–2): 267–277.

Goethe, J.W. (1980) *Le Traité des couleurs*. Paris: Éditions Triades.

Goh, C.S., D. Milburn, M. Gerstein (2004) Conformational Changes Associated with protein-protein interactions, *Current opinion in structural biology*, 14(1): 104–109.

Goodsell, D.S. (2003) Looking at molecules: An essay on art and science, *ChemBioChem*, 4: 1293–1298.

Gourdon, C., V. Jeudy, A. Cēbers, A. Dourlat, Kh. Khazen, A. Lemaître (2009) Unusual domain-wall motion in ferromagnetic semiconductor films with tetragonal anisotropy, *Physical Review B*, 80(16): 1202.

Granek, G. Hon (2008) Searching for asses, finding a kingdom: The story of the invention of the Scanning Tunneling Microscope (STM), *Annals of Science*, 65(1): 101–125.

Granqvist, F., R.A. Buhrman (1976) Ultrafine metal particles, *Journal of Applied Physics*, 47: 220–2219.

Greenleaf, W.J., M.T. Woodside, S.M. Block (2007) High-resolution, single-molecule measurements of biomolecular motion, Annual *Review of Biophysics and Biomolecular Structure*. Book Series: *Annual Review of Biophysics*, 36: 171–190.

Grüne-Yanoff, T., P. Weirich (2010) The philosophy and epistemology of simulation: A review, *Simulation & Gaming*, 41(1): 20–50.

Guzelian, A.A., U. Banin, A.V. Kadavanich, X. Peng, A.P. Alivisatos (1996) Colloidal chemical synthesis and characterization of InAs nanocrystal quantum dots, *Applied Physics Letters*, 69.

Ha, T., D.S. Chemla, T. Enderle, S. Weiss (1997) Single molecule spectroscopy with automated positioning, *Applied Physics Letters*, 70(6): 782–784.

Ha, T., T. Enderle, P.R. Selvin, D.S. Chemla, S. Weiss (1997) Quantum jumps of single molecules at room temperature, *Chemical Physics Letters*, 27(1–3): 1–3.

Ha, T., A.Y. Ting, J. Liang, W.B. Caldwell, A.A. Deniz, D.S. Chemla, P.G. Schultz, S. Weiss (1999) Single-molecule fluorescence spectroscopy of enzyme conformational dynamics and cleavage mechanism, *Proceedings of the National Academy of Sciences USA*, 96(3): 893–898.

Ha, T., A.Y. Ting, J. Liang, A.A. Deniz, D.S. Chemla, P.G. Schultz, S. Weiss (1999) Temporal fluctuations of fluorescence resonance energy transfer between two dyes conjugated to a single protein, *Chemical Physics*, 247(1): 107–118.

Haber, L.F. (1986) *The Poisonous Cloud: Chemical Warfare in the First World War*. Oxford: Clarendon Press.

Hacking, I. (2006) "An other world is being constructed right now: The ultra-cold." Conference *"The Shape of Experiment Berlin,"* 2–5 June, Max Planck Institute for the History of Science.

Hahn, C.K., T. Sugaya, K.Y. Jang, X.L. Wang, M. Ogura (2003) Electron transport properties in a GaAs/AlGaAs quantum wire grown on V-grooved GaAs substrate by metalorganic vapor phase epitaxy, *Japanese Journal of Applied Physics Part 1–Regular Papers Short Notes & Review Papers*, 42(4b): 2399–2403.

Harley, J.B., D. Woodward (eds.) (1987–1998) *The History of Cartography*, Volume 2, Book 1: *Cartography in the Traditional Islamic and South Asian Societies*. Chicago, IL/London: University of Chicago Press (2 volumes).

Harper, P.T.F., J.D. Wong, S.S. Lieber, C.M. Lansbury (1999) Assembly of A beta amyloid protofibrils: An in vitro model for a possible early event in Alzheimer's disease, *Biochemistry*, 38(28): 8972–8980.

Heath, J.R., M.E. Davis, L. Hood (2009) Nanomedicine targets cancer, *Scientific American*, 300(2): 44–51.

Hellinga, H. (1997/2004) Biosensor is a model for widespread testing of chemicals, molecules, *Duke Medicine News and Communications*.

Hennig, J. (2004) Changes in the design of scanning tunneling microscopic images from 1980 to 1990, *Techné: Research in Philosophy and Technology*, 8(2).

Hesse, M.B. (1966) *Models and Analogies in Science*. Notre Dame, IN: University of Notre Dame Press.

Hoddeson, L., E. Braun, J. Teichmann, S. Weart (eds.) (1992) *Out of the Crystal Maze: Chapters from the History of Solid-State Physics*. Oxford: Oxford University Press.

Hou, J.G., X.D. Xiang, M. Cohen, A. Zettl (1994) Granularity and upper critical fields in K3C60, *Physica C*, 232(1–2): 22–26.

Hou, J.G., X.D. Xiang, V.H. Crespi, M.L. Cohen, A. Zettl (1994) Magnetotransport in single-crystal Rb3Cc60, *Physica C*, 228(1–2): 175–180.

Hubble, E.P. (1936a) Extra-galactic nebulae, *Contributions from the Mount Wilson Observatory / Carnegie Institution of Washington*, 324: 1–49.

Hubble, E.P. (1936b) *The Realm of the Nebulae*. New Haven, CN: Yale University Press.

Humphrey, P. (1995) Computational science and scientific method, *Minds and Machines*, 5(4): 499–512.

Humphrey, P. (2004) Extending ourselves, science at century's end, *Philosophical Questions on the Progress and Limits of Science*, 13.

Humphrey, P. (2009) The philosophical novelty of computer simulation methods, *Synthese*, 169(3): 615–626.

Humphrey, P., C. Imbert (eds.) (2012) *Models, Simulations and Representations*. London: Routledge.

Humphrey, W., A. Dalke, K. Schulten (1996) VMD: Visual molecular dynamics, *Journal of Molecular Graphics*, 14(1): 33–41.

Hussain L., S. von Bardeleben, H. Bernard, M. Lemaître, A. Gourdon, C. (2012) High domain wall velocities in in-plane magnetized (Ga,Mn)(As,P) layers, Thevenard, *Physical Review B*, 85(6): 4419.

Huynh, A., N.D. Lanzillotti-Kimura, B. Jusserand, B. Perrin, A. Fainstein, M.F. Pascual-Winter, E. Peronne, A. Lemaitre (2006) Subterahertz phonon dynamics in acoustic nanocavities, *Physical Review Letters*, 97(11).

Huynh, A., B. Perrin, B. Jusserand, A. Lemaitre (2011) Terahertz coherent acoustic experiments with semiconductor superlattices, *Applied Physics Letters*, 99(19).

Iijima, S. (1991) Helical microtubules of graphitic carbon, *Nature*, 354(6348): 56–58.

Iyer, G., F. Pinaud, J. Xu, Y. Ebenstein, J. Li, J. Chang, M. Dahan, S. Weiss (2011) Aromatic aldehyde and hydrazine activated peptide coated quantum dots for easy bioconjugation and live cell imaging, *Bioconjugate Chemistry*, 22(6): 1006–1011.

Jacob F. (1970) *La Logique du Vivant. Une Histoire de l'Hérédité*. Paris: Gallimard.

Jacob F. (1977) Evolution and tinkering, *Science*, 196(4295): 1161–1166.

Jacob F. (1987) *La Statue Intérieure*. Odile Jacob éditions.

Jacob, P. (1980) *De Vienne à Cambridge*. Paris: Gallimard.

Jacobs, J.A. (2013) *In Defense of Disciplines. Interdisciplinarity and Specialization in the Research University*. Chicago: The University of Chicago Press.

Jammer, M. (1966) *The Conceptual Development of Quantum Mechanics*. New York, NY: McGraw-Hill.

Jammer, M. (1974) *The Philosophy of Quantum Mechanics: The Interpretations of Quantum Mechanics in Historical Perspective*. New York, NY: Wiley-Interscience.

Joachim, C. (1987) Control of the quantum path target state distance – Bistable-like characteristic in a small tight-binding system, *Journal of Physics A–Mathematical and General*, 20(17): L 1149–L 1155.

Joachim, C. (1988) Molecular switch – A tight-binding approach, *Journal of Molecular Electronics*, 4(2): 125–136.

Joachim, C., J.P. Launay (1984) The possibility of signal molecular processing, *Nouveau Journal de Chimie – New Journal of Chemistry*, 8(12): 723–728.

Joachim, C., J.P. Launay (1986) Bloch effective Hamiltonian for the possibility of molecular switching in the ruthenium bipyridylbutadiene ruthenium system, *Chemical Physics*, 109(1): 93–99.

Joachim, C., H. Tang, F. Moresco, G. Rapenne, G. Meyer (2002) The design of a nanoscale molecular barrow, *Nanotechnology*, 13(3): 330–335.

Joachim, C., L. Plévert (2008) *Nanosciences. La révolution invisible*. Paris: Editions du Seuil.

Joerges, B., T. Shinn (2002) *Instrumentation between Science, State and Industry*. Dordricht: Springer.

Johnson, A. (2006) The shape of molecules to come. In: G. Küppers, J. Lenhard, T. Shinn (eds.) *Simulation: Pragmatic Construction of Reality*. Dordrecht: Springer, pp. 25–39.

Jotterand, F. (ed.) (2008) *Emerging Conceptual, Ethical and Policy Issues in Bionanotechnology*. Dordrecht: Springer.

Kay, L.E. (1993) *The Molecular Vision of Life: Caltech, the Rockefeller Foundation, and the Rise of the New Biology*. New York, NY: Oxford University Press.

Kay, L.E. (2000) *Who Wrote the Book of Life? A History of the Genetic Code*. Stanford, CA: Stanford University Press.

Keller, E.F. (1990) Physics and the emergence of molecular biology: A history of cognitive and political synergy, *Journal of the History of Biology*, 23(3), 389–409.

Kish, L.B., J. Söderlund, G.A. Niklasson, C.G. Granqvist (1999) New approach to the origin of lognormal size distributions of nanoparticles, *Nanotechnology*, 10: 25–28.

Klein, D., R. Roth A. Lim, P. Alivisatos, L. McEuen (1997) A single-electron transistor made from a cadmium selenide nanocrystal, *Nature*, 389: 699.

Kohler, R.E. (2002) *Landscapes and Labscapes: Exploring the Lab–field Border in Biology*. Chicago, IL / London: University of Chicago Press.

Kroto, H.W., R.E. Smalley, J.R. Heath, R.F. Curl (1985) C60: Buckminsterfullerene, *Nature*, 318(6042): 162–163.

Küppers, G., J. Lenhard (2006) Simulation and a revolution in modelling style: From hierarchical to network-like integration. In: G. Küppers, J. Lenhard, T. Shinn (eds.) *Simulation: Pragmatic Constructions of Reality*. Dordrecht: Springer, pp. 89–106.

Küppers, G., J. Lehnard, T. Shinn (eds.) (2006) *Simulation: Pragmatic Construction of Reality*. Dortrecht: Springer.

Lacharmoise, P., A. Fainstein, B. Jusserand, B. Perrin (2004) Semiconductor phonon cavities, *11th International Conference on Phonon Scattering in Condensed Matter, Proceedings*, pp. 2698–2701.

Lamont, M., G. Mallard, J. Guetzkow (2006) Beyond blind faith: Overcoming the obstacles to interdisciplinary evaluation, *Research Evaluation*, 15(1): 43–55.

Lanzillotti-Kimura, N.D., A. Fainstein, B. Perrin, B. Jusserand, O. Mauguin, L. Largeau, A. Lemaitre (2010) Bloch oscillations of THz acoustic phonons in coupled nanocavity structures, *Physical Review Letters*, 104(19).

Lastapis, M., M. Martin, D. Riedel, L. Hellner, G. Comtet, G. Dujardin (2005) Picometer-scale electronic control of molecular dynamics inside a single molecule, *Science*, 308(5724): 1000–1003.

Latour, B., S. Woolgar (1986) *Laboratory Life: The Construction of Scientific Facts*. Princeton, NJ: Princeton University Press.

A. Lattes (2000) De l'hydrogénation catalytique à la théorie chimique de la catalyse: Paul Sabatier, chimiste de génie, apôtre de la décentralisation, *Comptes Rendus de l'Académie des Sciences-Series IIC-Chemistry*, 3(9): 705–709.

Lawrence, J., T. Funkhouser (2004) A painting interface for interactive surface deformations, *Graphical models*, 66(6): 418–438.

Lécuyer, C. (2007) *Making Silicon Valley. Innovation and the Growth of High Tech*, 1930–1970. Cambridge, MA: MIT Press.

Lécuyer, C., D. Brock (2010) *Makers of the Microchip. A Documentary History of the Fairchild Semiconductor*. Cambridge, MA: MIT Press.

Lee, K.B., S. Park, C.A. Mirkin (2004) Multicomponent magnetic nanorods for biomolecular separations, *Angewandte Chemie–International Edition*, 43(23): 3048–3050.

Lee, S.W., B.K. Oh, R.G. Sanedrin, K. Salaita, T. Fujigaya, C.A. Mirkin (2006) Biologically active protein nanoarrays generated using parallel dip-pen nanolithography, *Advanced Materials*, 18(9): 1133.

Lee, Y.G., K.W. Lyons, S.C. Feng (2004) Software architecture for a virtual environment for nano scale assembly (VENSA), *Journal of Research of the National Institute of Standards and Technology*, 109(2): 279–290.

Lemaine, G., M. Mulkay, P. Weingart (1976) *Perspectives on the Emergence of Scientific Disciplines*. Paris: Mouton.

Lenhard, J. (2004) Nanoscience and the janus-faced character of simulations. In: D. Baird, A. Nordmann, J. Schummer (eds.) (2004) *Discovering the Nanoscale*. Landsdale, PA: IOS Press Inc., pp. 93–100.

Lenhard, J. (2007) Computer simulation: The cooperation between experimenting and modeling, *Philosophy of Science*, 74(2): 176–194.

Lenhard, J. (2010) Computation and simulation. In: J.T. Klein, C. Mitcham, R. Frodeman (eds.) *The Oxford Handbook of Interdisciplinarity*. Oxford: Oxford University Press.

Lenhard, J. (forth coming) Disciplines, models and computers. The path to computational chemistry, *Studies in History and Philosophy of Modern Physics*.

Lenhard, J., G. Küppers, T. Shinn (eds.) (2006) *Simulation: Pragmatic Constructions of Reality*. Dortrecht: Springer.

Lenoir, T. (1997) *Instituting Science. The Cultural Production of Scientific Disciplines*. Stanford, CA: Stanford University Press.

Leydesdorf, L., P. Zhou (2007) Nanotechnology as a field of science: Its delineation in terms of journals and patents, *Scientometrics*, 70(3): 693–713.

Lian, Y.C., J.C.M. Li (1992) A nanocavity in a FCC crystal, *Materials Chemistry and Physics*, 32(1): 87–94.

Linnaeus, C. (1758) *Systema Naturae*. Stockholm: Laurentius Salvius.

Loksztejn, A., Z. Scholl, P.E. Marszalek (2012) Atomic force microscopy captures ribosome bound nascent chains, *Chemical Communications*, 48(96): 11727–11729.

Loeve, S. (2009) *Le Concept de technologie à l'échelle des molécules-machines. Philosophie des techniques à l'usage des citoyens du nanomonde*. Thèse de Doctorat de philosophie, d'épistémologie et d'histoire des sciences et des techniques.niversité de Paris-Ouest.

Lynch, M. (1985) *Art and Artifact in Laboratory Science: A Study of Shop Work and Shop Talk in a Research Laboratory*. London: Routledge & Kegan Paul.

Lynch, M. (1990) The externalized retina: Selection and mathematization in the visual documentation of objects in the life sciences. In: M. Lynch, S. Woolgar (eds.) *Representation in Scientific Practice*. Cambridge, MA, London: MIT Press, pp. 153–186.

Lynch, M. (2006a) The production of scientific images. Visions and re-vision in the history, philosophy and sociology of science. In: L. Pauwels (ed.) *Visual Cultures of Science: Rethinking Representational Practices in Knowledge Building and Science Communication*. Dartmouth, NH: Dartmouth College Press, pp. 26–41.

Lynch, M. (2006b) Discipline and the material form of images. An analysis of scientific visibility. In: L. Pauwels (ed.) *Visual Cultures of Science: Rethinking Representational Practices in Knowledge Building and Science Communication*. Dartmouth, NH: Dartmouth College Press, pp. 195–232.

Lynch, M., S.Y. Edgerton (1988) Aesthetic and digital image processing representational craft in contemporary astronomy. In: G. Fyfe, J. Law (eds.) *Picturing Power; Visual Depictions and Social Relations*. London: Routledge, pp. 184–220.

Lynch, M., S. Woolgar (eds.) (1990) *Representation in Scientific Practice*. Cambridge, MA: MIT Press.

Manchester, K.L. (2002) Man of destiny: The life and work of Fritz Haber, *Endeavour*, 26(2), 64–69.

Maoz, R., J. Sagiv (1984) On the formation and structure of self-assembling monolayers, 1: A comparative ATR-Wetability study of Langmuir-Blodgett and absorbed films on flat substrates and glass microbeads, *Journal of Colloid and Interface Science*, 100(2): 465–496.

Marangolo, M., F. Finocchi (2008) Fe-induced spin-polarized electronic states in a realistic semiconductor tunnel barrier, *Physical Review B*, 77(11).

Marcovich, A. (1976) La Créativité des hommes de sciences, Diplome de l'École des Hautes Études en Sciences Sociales, Paris.

Marcovich, A., T. Shinn (2010) Socio/intellectual patterns in nanoscale research. Feynman Nanotechnology Prize laureates, 1993–2007, *Social Science Information*, 49(4): 615–638.

Marcovich, A., T. Shinn (2011a) Instrument research, tools, and the knowledge enterprise 1999–2009 Birth and development of Dip-Pen nanolithography, *Science, Technology & Human Values*, 36(6): 864–896.

Marcovich, A., T. Shinn (2011b) The cognitive, instrumental and institutional origins of nanoscale research: The place of biology. In: *Science in the Context of Application*. Netherlands: Springer, pp. 221–242.

Marcovich, A., T. Shinn (2011c) Estrutura e funçao das imagens na ciência e na arte: entre a sintese e o holismo da forma, da força e da perturbaçao, *Scientiae Studia*, 9(2): 229–267.

Marcovich, A., T. Shinn (2011d) From the triple helix to a quadruple helix? The case of Dip-Pen nanolithography, *Minerva*, 49(2): 175–190.

Marcovich, A., T. Shinn (2011e) Where is disciplinarity going? Meeting on the borderland, *Social Science Information*, 50(3–4): 582–606.

Marcovich, A., T. Shinn (2013a) Respiration and cognitive synergy: Circulation in and between scientific research spheres, *Minerva*, 1(51): 1–23.

Marcovich, A., T. Shinn (2013b) Computer simulation and the growth of nanoscale research in biology. In: J. Duran, E. Arnold (eds.) *Computer Simulation and the Changing Face of Scientific Experimentation*. New Castle: Cambridge Scholars Publishing, pp. 145–170.

Marr, D. (1976) Early processing of visual information, *Philosophical Transactions of the Royal Society of London. Series B. Biological Sciences*, 275(942): 483–519.

Marr, D., H.K. Nishihara (1978) Representation and recognition of the spatial organization of three-dimensional shapes, *Proceedings of the Royal Society of London. Series B. Biological Sciences*, 200(1140): 269–294.

Matricon, J., G. Waysand (2003) *The Cold Wars: A History of Superconductivity*. New Bruswick, NJ: Rutgers University Press.

McCabe, D.P., A.D. Castel (2008) Seeing is believing: The effect of brain images on judgments of scientific reasoning, *Cognition*, 107(1): 343–352.

McCray, P. (2007) MBE deserves a place in the history books, *Nature Nanotechnology*, 2: 259–261.

McCray, P. (2012) *The Visioneers: How a Group of Elite Scientists Pursued Space Colonies, Nanotechnologies, and a Limitless Future*. Princeton, NJ: Princeton University Press.

Meyer, M., O. Persson (1998) Nanotechnology–interdisciplinarity, patterns of collaboration and differences in application, *Scientometrics*, 42(2): 195–205.

Milburn, C. (2010) *Nanovision: Engineering the Future*. Durham, NC: Duke University Press Books.

Mody, C. (2004) Small but determined: Technological determinism in nanoscience, *Hyle-International Journal of Philosophy of Chemistry*, 10(2): 99–128.

Mody, C. (2006) Corporations, universities, and instrumental communities: Commercializing probe microscopy, 1981–1996, *Technology and Culture*, 47(1): 56–80.

Mody, C. (2011) *Instrumental Community: Probe Microscopy and the Path to Nanotechnology*. Cambridge, MA: MIT Press.

Mody, C., M. Lynch (2010) Test objects and other epistemic things: A history of a nanoscale object, *The British Journal of The History of Science*, 43(3): 423–458.

Mohebi, A., P. Fieguth (2006) Posterior sampling of scientific images. In: A. Campilho, M.S. Kamel (eds.) *Image Analysis and Recognition*. Berlin, Heidelberg: Springer, pp. 339–350.

Molinas-Mata, P., A.J. Mayne, G. Dujardin (1998) Manipulation and dynamics at the atomic scale: A dual use of the scanning tunneling microscopy, *Physical Review Letters*, 80(14): 3101–3104.

Morange, M. (1998) *A History of Molecular Biology*. Cambridge, MA: Harvard University Press.

Morgan, M.S., M. Morrison (eds.) (1999) *Models as Mediators: Perspectives on Natural and Social Science*. Cambridge: Cambridge University Press.

Mulkay, M. (1974) Conceptual displacement and migration in science: A preparatory paper, *Science Studies*, 4: 205–234.

Mullins, N. (1972) The development of a scientific specialty: The phage group and the origins of molecular biology, *Minerva*, 10(1): 51–82.

Noguera, C., F. Finocchi, J. Goniakowski (2004) First principles studies of complex oxide surfaces and interfaces, *Journal of Physics-Condensed Matter*, 16(26): 2509–2537.

Novotny, H. (2007) The history of near-field optics. In: E. Wolf (ed.) *Progress in Optics*. Amsterdam: Elsevier, pp. 137–184.

Nowotny, H. (with P. Scott, M. Gibbons) (2001) *Re-Thinking Science. Knowledge and the Public in an Age of Uncertainty*. Oxford: Polity.

Nowotny, H., U. Felt (1997) *After the Breakthrough: The Emergence of High Temperature Superconductivity as a Research field*. Cambridge: Cambridge University Press.

Nye, M.J. (1972) *Molecular Reality: A Perspective on the Scientific Work of Jean Perrin*. London: Macdonald.

Nye, M.J. (1994) *From Chemical Philosophy to Theoretical Chemistry: Dynamics of Matter and Dynamics of Disciplines, 1800–1950*. Berkley, CA: University of California Press.

Ozin, G. (1992) Nanochemistry: Synthesis in diminishing dimensions, *Advanced Materials*, 4(10): 612–649.

Pan, J., Y. Wang, S.-S. Feng (2008) Formulation, characterization, and in vitro evaluation of quantum dots loaded in poly(lactide)-vitamin E TPGS nanoparticles for cellular and molecular imaging, *Biotechnology and Bioengineering*, 101(3): 622– 633.

Park, J.W., A.-Y. Park, S. Lee, N.-K. Yu, S.-H. Lee, B.-K. Kaang (2010) Detection of TrkB receptors distributed in cultured hippocampal neurons through bioconjugation between highly luminescent (quantum dot-Neutravidin) and (Biotinylated Anti-TrkB antibody) on neurons by combined atomic force microscope and confocal laser scanning microscope, *Bioconjugate Chemistry*, 21(4): 597–603.

Pauwels, L. (ed.) (2006) *Visual Cultures of Science: Rethinking Representational Practices in Knowledge Building and Science Communication*. Dartmouth, NH: Dartmouth College Press.

Pei, R., A.L. Porter (2011) Profiling leading scientists in nanobiomedical science: Interdisciplinarity and potential leading indicators of research directions, *R&D Management*, 41(3): 288–306.

Pelesko, J.A. (2007) *Self-Assembly: The Science of Things that Put Themselves Together*. New York, NY: Chapman & Hall/CRC.

Peng, W., F. Jabeen, B. Jusserand, J.C. Harmand, M. Bernard (2012) Conduction band structure in wurtzite GaAs nanowires: A resonant Raman scattering study, *Applied Physics Letters*, 100(7).

Piner, R.D., J. Zhu, F. Xu, S.H. Hong, C.A. Mirkin (1999) "Dip–pen" nanolithography, *Science*, 283(5402): 661–663.

Pitt, J. (2005) When is an image not an image, *Techné: Research in Philosophy and Technology*, 11(1).

Porter, A.L. (to appear *R&D*) Management profiling leading scientists in nano-biomedical science: Interdisciplinarity and potential leading indicators of research directions; "Ruimin Pei Chinese Academy of Sciences School of Public Policy Georgia Institute of Technology"

Porter, A.L, I. Rafols (2009) Is science becoming more interdisciplinary? Measuring and mapping six research fields over time, *Scientometrics*, 81(3): 719–745.

Porter, A.L., J. Youtie (2009) How interdisciplinary is nanotechnology? *Journal of Nanoparticle Research*, 11(5): 1023–1041.

Rabinow, P. (1996) *Making PCR. A Story of Biotechnology*. Chicago, IL: University of Chicago Press.

Rafols, I. (2007) Strategies for knowledge acquisition in bio-nanotechnology: Why are interdisciplinary practices less widespread than expected? *Innovation: The European Journal of Social Science Research*, 20(4): 395–412.

Rafols, I., M. Meyer (2007) How cross-disciplinary is bionanotechnology? Explorations in the specialty of molecular motors, *Scientometrics*, 70(3): 633–650.

Rafols, I., M. Meyer (2010) Diversity and network coherence as indicators of interdisciplinarity: Case studies in bionanoscience, *Scientometrics*, 82(2): 263–287.

Rawls, J., R.J. Orsi (eds.) (1999) *A Golden State: Mining and Economic Development in Gold Rush California*. Berkeley / Los Angeles, CA: University of California Press (California History Sesquicentennial, 2).

Reed, J., C. Hsueh, M.L. Lam, R. Kjolby, A. Sundstrom, B. Mishra, J. Gimzewski (2012) Identifying individual DNA species in a complex mixture by precisely measuring the spacing between nicking restriction enzymes with atomic force microscope, *Journal of the Royal Society Interface*, 9(74): 2341–2350.

Reed, J., S. Ramakrishnan, J. Schmit, J. Gimzewski (2009) Mechanical interferometry of nanoscale motion and local mechanical properties of living zebrafish embryos, *ACS Nano*, 3(8): 2090–2094.

Reed, M. (1993) Quantum dots. Nanotechnologists can now confine electrons to pointlike structures. Such "designer atoms" may lead to new electronic and optical devices, *Scientific American*, 268(1): 118–123.

Reich, L.S. (1983). Irving Langmuir and the pursuit of science and technology in the corporate environment, *Technology and Culture*, 24(2): 199–221.

Reinhardt, C. (2006) *Shifting and Rearranging. Physical Methods and the Transformation of Modern Chemistry*. Sagamore Beach, MA: Science History Publications.

Reuss, A. (1929) Account of the liquid limit of mixed crystals on the basis of the plasticity condition for single crystal, *Zeitschrift für Angewandte Mathematik und Mechanik*, 9: 49–58.

Reynolds, J., C. Tanford (2004) *Nature's Robots: A History of Proteins*. Oxford: Oxford University Press.

Rheinberger, H.-J. (1997) *Towards a History of Epistemic Things: Synthesizing Proteins in the Test Tube*. Stanford, CA: Stanford University Press.

Riordan, M., L. Hoddeson (1997) *Crystal Fire: The Birth of the Information Age*. New York, NY: WW Norton & Company.

Rip, A. (2010) Visualizing the invisible nanoscale. Study of visualization practices in nanotechnology community of practice, *Science Studies*, 23(1): 3–36.

Robinson, C. (2004) Images in nanoscience / technology. In: D. Baird, A. Nordmann, J. Schummer (eds.) *Discovering the Nanoscale*. Amsterdam: IOS Press, pp. 165–169.

Roco, M.C. (2003) Nanotechnology: Convergence with modern biology and medicine, *Current Opinion in Biotechnology*, 14(3): 337–346.

Roco, M.C. (2004) Nanoscale science and engineering: Unifying and transforming tools, *AIChE J.*, 50(5): 890–897.

Roco, M.C., W.S. Bainbridge (2003) *Converging technologies for improving human performance: Nanotechnology, biotechnology, information technology and cognitive science*. Dordrecht: Kluwer Academic Publishers.

Rohrich, F. (1990) Computer simulation in the physical sciences, *Proceedings of the Biennial Meeting of the Philosophy of Science Association*, 2: 507–518.

Rong, W.Z., A.E. Pelling, A. Ryan, J. Gimzewski, S.K. Friedlander (2004) Complementary TEM and AFM force spectroscopy to characterize the nanomechanical properties of nanoparticle chain aggregates, *Nano Letters*, 4(11): 2287–2292.

Rothbart, D., J. Schreifels (2006) Visualizing instrumental techniques of surface chemistry. In: D. Baird, L.C. McIntyre, E.R. Scerri (eds.) *Philosophy of Chemistry: Synthesis of a New Discipline*. Dordrecht: Springer, pp. 309–324.

Rothemund, P.W.K. (2006) Folding DNA to create nanoscale shapes and patterns, *Nature*, 440(7082): 297–302.

Ruello, P., S. Zhang, P. Laffez, B. Perrin, V. Gusev (2009) Laser-induced coherent acoustical phonons mechanisms in the metal-insulator transition compound NdNiO3: Thermal and nonthermal processes, *Physical Review B*, 79(9).

Ruivenkamp, M. (2011) *Circulating Images of Nanotechnology*, Doctoral Dissertation, Universiteit Twente.

Ruivenkamp, M., A. Rip (2010) Visualizing the invisible nanoscale. Study of visualization practices in nanotechnology community of practice, *Science Studies*, 23(1): 3–36.

Sass, J.K., J. Gimzewski (1988) Proposal for the simulation of electrochemical charge-transfer in the scanning tunneling microscope, *Journal of Electroanalytical Chemistry*, 251(1): 241–245.

Schummer, J. (2004) Multidisciplinarity, interdisciplinarity, and patterns of research collaboration in nanoscience and nanotechnology, *Scientometrics*, 59(3): 425–465.

Seager, C., S.M. Myers, R.A. Anderson, W.L. Warren, D.M. Follstaedt (1994) Electrical-properties of He-implantation-produced nanocavities in silicon, *Physical Review B*, 50(4): 2458–2473.

Seeman, N.C. (2004) Nanotechnology and the double helix, *Scientific American*, 290(6): 64–75.

Semond, R., P. Soukiassian, A. Mayne, G. Dujardin, L. Douillard, C. Jaussaud (1996) Atomic structure of the beta-SiC(100)-(3x2) surface, *Physical Review Letters*, 77(10): 2013–2016.

Shannon, C. (1949) Communication theory of secrecy systems, *Bell System Technical Journal*, 28(4): 656–715.

Sharma, S.H., C. Santiskulvong, L.A. Bentolila, J.Y. Rao, O. Dorigo, J. Gimzewski (2012) Correlative nanomechanical profiling with super-resolution F-actin imaging reveals novel insights into mechanisms of cisplatin resistance in ovarian cancer cells, *Nanomedicine-Nanotechnology Biology and Medicine*, 8(5): 757–766.

Shepard, R.N. (1978) The mental image, *American Psychologist*, 33(2).

Shinn, T. (2006) When is simulation a research technology? Practice, markets and lingua franca. In: J. Kueppers, J. Lehnard, T. Shinn (eds.) *Simulation: Pragmatic Construction of Reality*. Dortrecht: Springer, pp.187–205.

Shinn, T. (2008) *Research-Technology and Cultural Change: Instrumentation, Genericity, Transversality*. Oxford: Bardwell.

Shinn, T. (2013) The silicon tide. In: R. Fox (ed.) *The Oxford Handbook of the History of Physics*. Oxford: Oxford University Press, pp. 860–892.

Sitti, M., S. Origuchi, H. Ashimoto (1998) Nano tele-manipulation using virtual reality interface, *IEEE Xplore*, 1(1): 171–176.

Smalley, R.E. (1996) *Discovering the Fullerenes*, Nobel Lecture, 7 December.

Soukiassian, P., F. Semond, L. Douillard, A. Mayne, G. Dujardin, L. Pizzagalli, C. Joachim (1997) Direct observation of a beta-SiC(100)-c(4x2) surface reconstruction, *Physical Review Letters*, 78(5): 907–910.

Staley, T.W. (2008) The coding of technical images of nanospace: Analogy, disanalogy, and the asymmetry of worlds, *Techne*, 12(1): 1–22.

Stranski, N. (1928) On the theory of crystal accretion, *Zeitschrift für Physikalische Chemie–Stochiometrie und Verwandtschaftslehre*, 136(3 / 4): 259–278.

Sturken, M., L. Cartwright (2001) *Practices of Looking; An Introduction to Visual Culture*. New York, NY: Oxford University Press.

Suppe, F. (1977) *The Structure of Scientific Theories*. Urbana, IL: University of Illinois Press.

Suppe, F. (1989) *The Semantic Conception of Theories and Scientific Realism*. Urbana, IL: University of Illinois.

Tanaka, K. (2003) The origin of macromolecule ionization by laser irradiation, Nobel lecture, 8 December 2002. *The Nobel Prizes 2002*, Stockholm: Tore Frängsmyr.

Thevenard, L., C. Gourdon, S. Haghgoo, J.-P. Adam, J. von Berdeleben, A. Lemaître, W. Schoch, A. Thiaville (2011) Domain wall propagation in ferromagnetic semiconductors: Beyond the one-dimensional model, *Physical Review B*, 83(24): 5211.

Thomson, R.E., U. Walter, E. Ganz, J. Clarke, A. Zettl, P. Rauch, F.J. Disalvo (1988) Local charge-density-wave structure in 1t-Tas2 determined by scanning tunneling microscopy, *Physical Review B*, 38(15): 10734–10743.

Thompson-Klein, J. (1991) *Interdisciplinarity. History, Theory and Practice*. Detroit, MI: Wayne State University Press.

Thompson Klein, J. (1996) *Crossing Boundaries: Knowledge, Disciplinarities, and Interdisciplinarities*. Charlottesville, VA/London: University Press of Virginia.

Thompson-Klein, J. (2005) *Humanities, Culture and Interdisciplinarity. The Changing American Academia*. New York, NY: State University of New York.

Toumey, C. (2007) Cubism at the nanoscale, *Nature Nanotechnology*, 2: 587–589.

Toumey, C. (2009) Truth and beauty at the nanoscale, *Leonardo*, 42(2): 151–155.

Toumey, C. (2010) Images and icons, *Nature Nanotechnology*, 5: 3–4.

Turner, S. (2000) What are disciplines? And how is interdisciplinarity different? In: P. Weingart, N. Sterh (eds.) *Practicing Interdisciplinarity*. Toronto: University of Toronto Press, pp. 46–65.

Urvoas, A., M. Valerio-Lepiniec, P. Minard (2012) Artificial proteins from combinatorials approaches, *Trends in Biotechnology*, 30(10): 512–520.

Vaucouleurs, G. (De) (1959) Classification and morphology of external galaxies, *Handbuch der Physik*, 53: 275.

Vegard, L. (1921) The constitution of the mixed crystals and the filling of space of the atoms, *Zeitschrift für Physik*, 5(April–July): 17–26.

Vesper, B.J., K. Salaita, H. Zong, C. Mirkin, A. Barrett, B. Hoffman (2004) Surface-bound porphyrazines: Controlling reduction potentials of self-assembled monolayers through molecular proximity/orientation to a metal surface, *Journal of the American Chemical Society*, 126(50): 16653–16658.

Voliotis, V., R. Grousson, P. Lavallard, E.L. Ivchenko, A.A. Kiselev, R. Planel (1993) Gamma-x mixing in type-ii GaAs/Alas short-period superlattices, *Journal de Physique IV*, 3(5): 237–240.

Voliotis, V., R. Grousson, P. Lavallard, E.L. Ivchenko, A.A. Kiselev, R. Planel (1994) Absorption-coefficient in type-ii GaAs/Alas short-period superlattices, *Physical Review B*, 49(4): 2576–2584.

Von Hippel, A. (1956) Molecular engineering, *Science*, 123: 315–317.

Von Hippel, A. et al. (ed.) (1959) *Molecular Science and Molecular Engineering*. Cambridge, MA: MIT Press.

Wang, X.L., S. Furue, M. Ogura, V. Voliotis, M. Ravaro, A. Enderlin, R. Grousson (2009) Ultrahigh spontaneous emission extraction efficiency induced by evanescent wave coupling, *Applied Physics Letters*, 94(9).

Wang, Y.H., et al. (2006) Controlling the shape, orientation, and linkage of carbon nanotube features with nano affinity templates, *Proceedings of the National Academy of Sciences USA*, 103(7): 2026–2031.

Weingart, P., N. Sterh (eds.) (2000) *Practicing Interdisciplinarity*. Toronto: University of Toronto Press.

Weiss, S., D.F. Ogletree, D. Botkin, M. Salmeron, D.S. Chemla (1993) Ultrafast scanning probe microscopy, *Applied Physics Letter*, 63(18): 2567–2569.

Weiss, S. (2000) Measuring conformational dynamics of biomolecules by single molecule fluorescence spectroscopy, *Natural Structural Biology*, 7(9): 724–729.

Weiss, S. (2006) Probing dynamic structures and molecular interactions at ultrahigh resolution in-vitro and in live cells using single molecules and single quantum dots. In: *The Ultrastructure of Life: A Symposium Opening the Centre for Ultrastructural Imaging at King's College London*. London.

Welsh, T., M. Ashikhmin, K. Mueller (2002) Transferring color to greyscale images, *Proceedings of the 29th Annual Conference on Computer Graphics and Interactive Techniques ACM Transactions on Graphics (TOG)–Proceedings of ACM SIGGRAPH*, 21(3): 277–280.

Wetlaufer, D.B. (1973) Nucleation, rapid folding, and globular intrachain regions in proteins, *Proceedings of the National Academy of Sciences USA*, 70(3): 697–701.

Whitley, R. (2000) *The Intellectual and Social Organization of the Sciences*. Oxford: Oxford University Press.

Wildoer, J.W.G., L.C. Venema, A.G. Rinzler, R.E. Smalley, C. Dekker (1998) Electronic structure of atomically resolved carbon nanotubes, *Nature*, 391(6662): 59–62.

Wilson, A.D. (1991) Mental representations and scientific knowledge: Boltzmann's Bild theory of knowledge in historical context, *Physis*, 28: 770–795.

Wilson L., P.T. Matsudaira, B. Ramaswamy, J.H. Horber (2002). *Atomic Force Microscopy in Cell Biology* (Vol. 68). Access Online via Elsevier.

Winsberg, E. (2006) Handshaking your way to the top: Simulation at the nanoscale, *Sociology of Sciences Yearbook*, 25(Part 3): 139–151.

Witkowski, J.A. (2005) *Inside Story: DNA to RNA to Protein*. New York, NY: Cold Spring Harbor Laboratory Press.

Wojnar, L. (1999) *Image Analysis; Applications in Materials Engineering*. London: CRC Press.

Wulff, G. (1901) On the question of speed of growth and dissolution of crystal surfaces, *Zeitschrift für Krystallographie und Mineralogie*, 34(5 / 6): 449–530.

Yoshie, T., A. Scherer, J. Hendrickson, G. Khitrova, H.M. Gibbs, G. Rupper, C. Ell, O.B. Shchekin, D.G. Deppe (2004) Vacuum Rabi splitting with a single quantum dot in a photonic crystal nanocavity, *Nature*, 432(7014): 200–203.

Zettl, A., G. Grunier (1983) Charge-density-wave transport in orthorhombic Tas3 0.3. Narrow-band noise, *Physical Review B*, 28(4): 2091–2103.

Zou, S., D. Maspoch, Y. Wang, C.A. Mirkin, G.C. Schatz (2007) Rings of single-walled carbon nanotubes: Molecular-template directed assembly and Monte Carlo modelling, *Nano Letters*, 7(2): 276–280.

INDEX